FOCUS ON PLUTO

by

Virginia Elenbaas

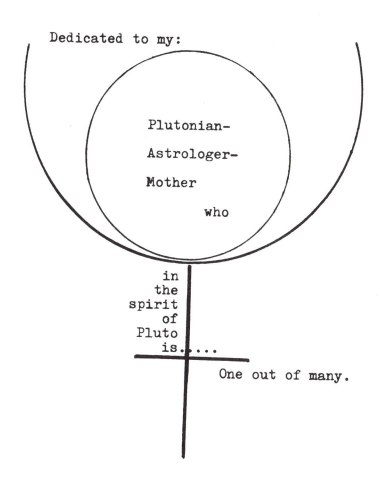

Dedicated to my:

Plutonian-

Astrologer-

Mother

who

in
the
spirit
of
Pluto
is.....

One out of many.

First Printing 1974
Third Printing 1985
ISBN Number: 0-86690-096-9

Published by:
American Federation of Astrologers, Inc.
P.O. Box 22040, 6535 South Rural Road
Tempe, Arizona 85282

Printed in the United States of America

INTRODUCTION

Some forty-odd years after its discovery in 1930, Pluto remains elusive and mysterious in the minds of many students of Astrology. It seems to be the big question mark...a blank spot in the horoscope, vaguely related to "group involvement", "regeneration", etc. Such terminology, while perhaps accurate, fosters interpretations which are long on confusion and short on usefulness.

This then, is an attempt to examine traditional beliefs, gain a new perspective by observing Pluto in action via the historical-national scene, and translate some of these observations into simple and utilitarian principles for natal interpretation.

Because Pluto's scope is broad, and its force powerfully enduring, it behooves us to understand the principles involved so that we may express the Pluto influence positively. To paraphrase a well-known quotation:[1]

> Living with Pluto is rather
> like a parachute jump....
> it's best to get it right
> the first time!

V.E.

TABLE OF CONTENTS

SECTION I

DATA

A compilation of accepted facts, figures and affinities pertinent to Pluto; a quick study for those who are new to the subject or a reference guide for those who are not.

DISCOVERY: January 21, 1930. Lowell Observatory, Flagstaff, Arizona

---

GLYPH: Represents the circle of potentiality over the semi-circle of receptivity, above the cross of materiality.

RULERSHIP:
Rules: Scorpio
Fall: Taurus
Dignified: Leo
Detriment: Aquarius

KEYWORDS:

Basic	Positive	Negative
Standardization	Transformation	Destruction
Transition	Miracles	Fanaticism
Barrier-Breaking	Universal Welfare	Regimentation
Reorganization	Rejuvenation	Atrocities
Extremism	Regeneration	Lawlessness
Group Action	Clairvoyance	Annihilation
Resurrection		
Force		
Intensification		
Anything Fantastic		
Obsession		
Investigation		
Prowess		

TRANSITS

Taurus:	1851..................................1884	
Gemini:	1882................................1914	
Cancer:	1912..........................1939	
Leo:	1937....................1958	
Virgo:	1956..............1972	
Libra:	1971...........1984	
Scorpio:	1984........1993	

PLANTS:

Pine
Cedar
Arborvitae
Cypress
Poplar
Willow
Mosses
Mushrooms
Maidenhair
Imortelle
Narcissus
Henbane
Witch's Bane
Poisonous growths
All plants with
    hairy leaves
Velvety plants and
    fruits; peaches,
    apricots
Coconuts
Herbs and plants
    producing heal-
    ing elixirs.

COLORS:

Dark purple-blue
Gentianella-blue
Reddish lilac
Brownish red
Yellowish brown
Sulphurous yellow
Dark grey, bluish tones
Black

GEMS:

Obsidian
Various dark crystals
Black diamonds
Pearls

WILD LIFE:

Scorpion
Lizard
Spider
Rat
Bat
Serpent
Eagle
Hyena
Vulture
Amphibians
Bacteria

OTHER:

Plutonium
Altars
Hearths
Pyramids
Old ruins
Grottoes
Damp cellars
Subterranean passages
Morgues
Haunted Houses
Vaults
Catacombs
Caves
Caverns
Cemeteries
Mausoleums
Graves
Sacrificial stones
Dark, quiet localities
Gloomy natural phenomena
Dark mountain gorges
Ravines
Curative muds
Streams
Mineral and hot springs
Marshy districts, swamps
Volcanic ash and lava
Stagnant water, ice
Decay, fermentation
Death, horror
Volcanoes, Earthquakes

SECTION II

THE CYCLES OF PLUTO

For a quick and clear understanding of Plutonian effects, there is no surer way than a backward glance through history. When judging personality traits of friends and family we all too often tend to see what we want or expect to see. Interpretations can vary widely when dealing with individuals, but one can hardly argue away World War II.

The question then is, did or did not certain events take place within the Plutonian context which were pertinent to the sign involved? What kind of force was present, and did the pattern repeat itself? If such is found, we can then relate this quality more objectively to individuals and also be more predictive about future events, a "must" for any science, since all true science is predictive.

We will begin about 100 years ago with the Pluto in Taurus Era, and follow Pluto's path up to the present, with the following qualifications:

1. Time of initial entry until final exit will be judged as the appropriate span of influence. Retrograde periods will be noted, but viewed in terms of both signs involved.

2. This study is limited to the United States because other countries may or may not relate in the same way. They would have to be studied individually.

National Implications

Since Pluto is considered ruler of mass movements, large scale destruction, and anything which is of a fantastic or extreme nature, we can expect to find these qualities demonstrated during each Pluto transit. Having already examined these eras and found this to be generally quite accurate, the premise can be stated accordingly:

1. Some form of annihilation or mass destruction which thus far has exhibited itself as war. With one exception, we have had exactly one war per Pluto period. Its"personality" and most often the direct causes are clearly linked to the sign through which Pluto is traveling.

3.

2. Some unusual <u>achievement</u>, usually in the interests of <u>universal welfare</u>; a manifestation of the extremist nature of Pluto and its ability to reach heights as well as depths.

3. A <u>middle turning point,</u> frequently accompanied by a crisis. This seems almost infallible, but I have no way to accurately pinpoint the date without more detailed calculations. It occurs approximately at the halfway mark, but does not always coincide with the actual midpoint.

4. We must remember that one can never think of just one sign, "Leo" or "Virgo", but must always think in terms of"Leo-Aquarius" or "Virgo-Pisces". For the sake of simplicity, I usually limit my wording to the primary sign, but both expressions should always be considered. After the middle turning point the polarity expression seems to emerge even more strongly, so we can then look for a <u>polarity</u> emphasis.

5. A tremendous surging forth of the qualities or expressions of the Pluto-related sign, resulting in some appropriate phenomenon in terms of <u>group</u> <u>activity</u> or group <u>movement</u>. Since large numbers of people are affected, it can frequently be seen through government, industry, education, etc., but is not necessarily limited to these areas. This is undoubtedly the most difficult area about which to be accurately predictive, since it will affect one aspect of the sign, leaving other equally likely aspects untouched. It can be easily recognized in retrospect, and its effects are always very enduring.

6. There are occasional significant occurrences which do not fit properly into our current framework of astrological understanding, and it seems sensible to face it squarely. Therefore, we may find some important <u>non-correlating</u> event.

4.

PLUTO IN TAURUS (SCORPIO):   June, 1851 - May, 1884

> "A nation may be said to consist of its territory,
> its people, and its laws. The territory is the
> only part which is of certain durability."
>
>   Abraham Lincoln [1].

Taurus-Aries Period: 1851,1852*
Taurus-Gemini Period: 1882,1883
Midpoint: 1867,1868

Venus-ruled, fixed-earth sign Taurus relates to the
land, industry, money, and all natural resources in
general. It is the essence of aggregativeness, building, constructive activity, and gathering-together.
Its expression is steady, stable, and sturdy. Keynote:
Doggedness.

Major Events and/or Issues:

The Civil War
Agricultural Emphasis
Emergence of U.S. as an Industrial Giant
Rise of Capitalism
Transcontinental Railroad System

1. Annihilation - Pluto in Taurus

    The Civil War: 1861-1865
    Slavery
    Free vs. Slave Labor

To appreciate the Taurean-related causes for this war
we must first set the stage. Out of 140,000 factories
in the U.S. at the time in question, 74,000 were lo-
cated in the Northeast. This was a thriving industrial
section with a hierarchy based on economic control and
industrial power.[2].

The Northwest was primarily agricultural and espoused
a system based on the democratic premise of equality.
Everyone had the right to acquire property in a fair

* Refers to retrograde period, not cusp reference.

                                                    5.

competition, and the notion prevailed that those with
ability and energy would automatically secure property
and those with special talents could rightfully acquire more.[3] Note that already the special ingredients
of our drama have strong Taurean overtones: free
enterprise, hard work, agricultural interests, etc.

The third party in what evolved as an unhappy triangle
was the South, one of the few areas in the whole world
still supporting slavery. Here, the planters were the
ruling hierarchy. It was an orderly, stable, conserva-
tive society wherein slaves were seen as the best solution to conflicts between capital and labor.[4] In this
instance, however, the solution was also the problem:
the South attracted none of the immigrant workers and
skilled craftsmen since they simply couldn't compete
in the system of slave labor. Growth was stymied and
the few crops produced created a dependent society.

There were, of course, men such as Emerson and other
humanitarians who objected to the institution of
slavery on moral grounds. Also, Harriet Beecher Stowe's
book, Uncle Tom's Cabin did arouse a certain amount of
public indignation, yet no move was made to interfere
in the southern system. The Supreme Court had even been
supportive through a decision in the Dred Scott Case
(1857) wherein they stated that "a Negro was the property of his owner, was not a citizen, and could never
be a citizen."[5] The very crux of slavery in fact, view-
ing a human being as one's "property", is well related
to negative Taurus.

The problem came to a head over the admission of new
territory, specifically Kansas. Both North and South
wanted to extend their special systems into the newly
settled land, yet free labor couldn't compete with
slave labor and so the tolerance which had existed
while slavery was contained within southern borders was
quickly abandoned and war broke out. Slavery was an
issue, but economics the cause....purely and simply a
Taurean case.

2. Achievement, Universal Welfare - Pluto in Taurus

Agriculture
Natural Resources
Industry

In the late 19th Century, farm workers far outnumbered

6.

all other occupations,[6] and this emphasis on "the land"
is very distinctly within the Taurean domain.

Great strides were made in agriculture. Before the
Civil War it took 61 hours to produce an acre of hand-
grown wheat; by the 1890's it took 3 hours, 19 minutes,
while the total production of corn and wheat doubled
between 1860 and 1880.[7] In 1884 the first commercial
flour mill was established in Minneapolis, and in 1856
Gail Borden patented his process for condensed milk.In
1862 the Federal Government established land grants
(Morrill Act of 1862) for Colleges of Agriculture and
the Mechanical Arts, and in that same year established
the Department of Agriculture. We have, of course, continued to improve our agricultural methods and all
other activities connected with the land in general, so
why the big fuss about this particular period? According to The Life History of the United States, this was
a period of "unparalleled expansion of America's natural resources with a burst of productive activity"[8]
(emphasis mine). Since Taurus is extremely productive
and rules "all natural resources", that would seem to
be close enough!

Man has been tilling the soil practically since the
dawn of time, and most of his tools had already been
developed before this particular era. In view of this
fact, the accomplishments of the Pluto in Taurus Era
become even more pertinent. Why not 50 years earlier,
or 50 years later? If studied from the astrological
point of view, there is no contest: Taurus rules the
land, agriculture, and hard labor, so this emphasis is
right on cue.

Although the U.S. was primarily an agrarian society,
industry too grew by leaps and bounds and by 1859 it
equalled agricultural production for the first time.
Since labor, heavy equipment, and the principle of
"building" are also Taurean, we're still on target.

With the beginnings of mass production, some ruthless
corporations surfaced, child labor problems emerged,
and in self defense, workers formed labor unions. Their
credo (quite Taurean) was: "An honest day's wage for an
honest day's work."[10]

By the end of the 19th Century, the world recognized
the U.S. as an industrial giant, and the businessman
was the most important figure in society.[11]

Whether we're discussing business, industry, or agri-

culture, the mass consciousness obviously leaned heavily in the Taurean direction of pure productive effort!

3. Middle Turning Point and/or Crisis - Pluto in Taurus

Midpoint Date: 1867, 1868

Reconstruction Period - 1866
14th Amendment Ratified - 1868

The Civil War began in 1861, Lincoln issued his famous
Emancipation Proclamation in 1862, and the war ended
in 1865; surely all significant dates which involved
earth-shattering changes, and thus overshadowed the
midpoint period crisis-wise. It was, in this instance,
a decided turning point with a concurrent swing to the
Scorpio expression. Consider the following:

1867: Three Reconstruction Acts passed. This is particularly apt for Scorpio, and is discussed further under
"Polarity Emphasis".

1868: President Johnson was impeached, tried and ac-
quitted; the only time this has happened in all our
history and correlates with militant Scorpio.

1868: The 14th Amendment to the Constitution was
passed, granting Black People citizenship.* This was
the final legal blow to slavery, since all the "freedom" granted by the Emancipation Proclamation and the
Civil War itself meant nothing without citizenship.

4. Polarity Emphasis - Pluto in Taurus

Polarity Sign: Scorpio   Ruler: Pluto

Qualities: "Throwing-out-ness", elimination, get-
            ting rid of waste, shedding the old.
            Its expression is passionate, strong,
            and determined. Keynote: Strength or
            Force

*While there are sometimes differences of opinion on
rulerships, the relationship of Scorpio to the dark
races seems valid to me. The Scorpio Eighth House is
associated with all that is hidden or "dark" while the
Scorpio type is usually described as dark or swarthy...
thus a logical link with all the dark races which includes but is of course not limited to the Black People.

8.

Events: The Reconstruction Era
           Passage of 14th and 15th Amendments

This post Civil War period is commonly called "The Reconstruction Era" which is very apt: while Taurus =
"construction", Scorpio = "reconstruction" and it is
interesting that historians would so accurately label
this time slot in terms of the astrological reference.
This remodeling process, which affected the entire
country, went into full swing in 1867 with the passage
of the Reconstruction Acts and was very appropriate in
terms of the polarity expression.

Also, as suggested previously, a strong Scorpio emphasis frequently brings some unusual degree of attention
to the dark races. The 14th (1868) and 15th (1870)
Amendments to the Constitution were passed during this
time redefining citizenship and granting the vote to
all persons regardless of race; all this, of course,
inspired by the needs of Black people.

In addition, the Scorpio proclivity for secretiveness
became a national phenomenon through the "underground
railroad;" a system whose object was to secretly assist
slaves into free territory. Barns and sheds were the
"stations" and the fugitive was passed along from station to station until he reached safety.

A less worthy but still significant event was the establishment of the Ku Klux Klan in 1866...a secret
order wherein the Klansmen wore shrouds symbolizing the
spirits of dead rebel soldiers; one of Scorpio's less
desirable offspring, to be sure, but relates well astrologically to "death" which is ruled by Scorpio.

5. Group Activity or Movement - Pluto in Taurus

    The Victorian Era
    Educational Rote and Routine
    Rise of Capitalism

The Pluto in Taurus period tallies quite closely with
what historians call "The Victorian Era", dated 1840-
1880. This is noted as a time of "rigid conformity" and
what could better describe solid, stolid Taurus? Neptune
was also transiting Taurus from 1875-1887, so that the
Taurean qualities are even more noticeable. In such instances it is quite possible to attribute to one planetary influence that which should be attributed to the
other so with that factor clearly in mind we shall

continue.

During the Victorian Era the Taurean attitude prevailed:
husbands were content to slave in business that their
families might live in ease and affluence. In the 1870's
the average breadwinner worked 10 hours a day, 6 days a
week.[12] This rather accurately reflects the Taurean approach to life which is essentially a practical, responsible one, and prone to "plug away" doggedly to achieve
a goal.

In the post Civil War period business boomed, and The
Boston Herald in 1900 wrote: "if a man could not have
made money this past year his case is hopeless." [13]
Prices and taxes were low while business flourished.
Could the situation possibly have been otherwise with
all that Taurean industriousness swamping the mass
consciousness?

The schools of this period were described as a "robot
parade ground".[14] Everything was drilled into young
minds by interminable repetition: Taurus in action. The
entire first school year, according to one approved
method of teaching math, was spent entirely on numbers
1 to 10, and again we can see the slow but sure Taurean
expression at work. Handwriting was a mechanistic night-
mare; drill books explained exactly where to place the
thumb, forefinger, etc. Public speaking was taught according to a precise, stylized set of movements which
the student was expected to memorize. Mark Twain described them as "the painfully exact and spasmodic ges-
tures which a machine might have used;" a fair depiction
of the Taurean expression (not necessarily the Taurean
person of course), which is usually predictable, conform-
ing and often rigid. This system of education is summed
up well as "rote and routine",[16] a descriptive title which
could have been coined from an astrological text on the
Taurean principle.

Finally, it was during this era that our industrious
money-oriented Pluto in Taurus consciousness brought
about the blossoming of the capitalistic system. Note
again that with the Pluto emphasis, systems and institutions are both born and wiped out; Pluto obviously
manifests on nothing less than a grand scale.

The "almighty dollar" was evidently on the national
mind, and the first president of Temple University,
Russell Conwell, shouted to students: "I say that you
ought to get rich, and it is your duty to get rich...
To make money honestly is to preach the gospel."[17]
Can anyone imagine such a philosophy voiced in today's

universities? The idea is quite outrageous, but was totally acceptable at the time, under the strong Taurean influence.

The Taurean principle of "building","accummulation", and "aggregation" brought huge monopolies into being with the consequent emergence of our first millionaires: Henry Clay Frick, J. Pierpont Morgan, Andrew Carnegie, John D. Rockefeller, etc.[18] Now we have reached the Taurean "big time". These people really understood monetary power and accumulation, and it surely wasn't entirely coincidental that they arrived on the scene with such precise timing, astrologically. Foundations bearing their names still exist some 100 years later; a living testimony to Pluto-inspired power.

Other significant developments which fit well into the Taurean-money context:

1849 - California Gold Rush (time-wise on the Aries-Taurus cusp, and a neat correlation with Aries impetuosity linked to Taurean "money")

1862 - The first Federal Income Tax

1862 - The first issue of greenbacks

1865 - Congress imposed a tax on state bank notes, forcing 700 state banks to close and ultimately bringing about a uniform currency.

1879 - The cash register invented

It becomes increasingly clear that Pluto, sometimes called "The Lonely Wanderer" is doing quite a bit besides wandering around!

6. Non-Correlating Events - Pluto in Taurus

Transcontinental Railroad System - 1869

With the exception of boats and wagons which go too far back to properly judge anyway, I can think of no other transportation forms which required such brute strength to construct as the railroads. The arduous labor in-volved and the heavy equipment of the railroad itself seem very Taurean, while our "travel" sign, Gemini is more typically "ingenuity".

The significance of this event is tremendous. Between 1860 and 1900, 60% of all the steel manufactured in the

U.S. went into rails, and by 1898 we had one-half of all the railway mileage in the world. By this time of course we are well into the Gemini period, noted as the "Golden Age of Railroads."[19]

In any case, this _is_ a form of transportation and must therefore fall into the Gemini domain. The date of completion was 13 years early when judged according to the Pluto reign and so must be judged inappropriate to the Pluto in Taurus Era, or at best, a borderline case.

7. Sidelights - Pluto in Taurus

*Uranus did tenant Taurus from 1851 - 1858, and Neptune from 1875-1887, yet only Pluto was transiting for the Civil War and Reconstruction Period. If we view the Civil War as appropriate to the Taurus-Scorpio frame of reference, then the significance of the Pluto involvement must be acknowledged.

*The most powerful organization for catching criminals in the late 19th Century was Pinkerton's National Detective Agency. Its motto was "We Never Sleep", and by 1870 it had become the world's largest agency of its type.[20] The Scorpio proclivity for "detecting" is highlighted here, and provides one more example of the polarity sign entrenching itself in our mode of life.

* Pluto fosters "fantastic" events, and one such example in this era was a court case, described by The National Enquirer (May 20, 1973) as "one of history's most colorful court battles". It concerned the contested will of millionaire Commodore Vanderbilt; one which had been drawn up according to advice from "spirits". The case dragged on for two years and featured a "parade of mediums, spiritualists, and bizarre testimony." Public attention was again focused on Taurus-Scorpio; money and the supernatural.

PLUTO IN GEMINI (SAGITTARIUS):   August,1882-June,1914

> "Education is a social process
> Education is growth
> Education is not preparation for life;
> Education is life itself."
>   John Dewey[1].

Gemini-Taurus Period: 1882,1883
Gemini-Cancer Period: 1912,1913
Midpoint: 1897,1898

Mercury-ruled, mutable air sign Gemini relates to com-
munications, transportation, and the ordinary, everyday affairs of life in general. It is the essence of
movement, change, linkage, and connectingness. Its
expression is buoyant, clever, and alert. Keynote:
Versatility.

Major Events and/or Issues

Spanish-American War
Advances in Transportation, Communication
Migration, Immigration, Urbanization

1. Annihilation - Pluto in Gemini

The Spanish-American War: 1898
Journalism Inspires "War Fever"

William Randolph Hearst and Joseph Pulitzer, the two
leading newspaper publishers in the 90's, were consis-
tently involved in a highly competitive campaign to cap-
ture customers. Spanish "atrocities" in Cuba made good
front-page news at the time, and consequently boosted
circulation. Headlines read: "Feeding prisoners to
sharks", or "Beating prisoners to death."[2] Day after day
they hammered out stories that our neighbors, the Cubans,
were being massacred by the evil Spanish. These New York
papers were soon copied throughout the country and even-
tually aroused the indignation of the entire American
public. President McKinley and his cabinet were not
anxious for war, but "war fever" seemed to be rising
both in Congress and the public in general; the newspapers had inspired a national hysteria.[3] Notice that
already we have two ingredients of a Gemini plot: concern
over neighbors, and the written word.

When the American ship, the "Maine" was sunk in a Cuban
harbor killing 260 men, the public reacted fiercely.
The Spanish sought to appease with apologies and ex-
planations, but American anger was not to be dismissed.
The exact cause to this day remains a mystery, but
"causes" were not considered at the time. Pulitzer and
Hearst papers screamed war headlines: "The Warship
Maine Was Split In Two By An Enemy's Secret Infernal
Machine!"...."Remember The Maine!"...."To Hell With
Spain!"[4] Shortly thereafter we marched off to our
"Splendid Little War."[5] The light-hearted Gemini spirit
prevailed even facing the horror of war and it was
said that we fought with "zest, clumsiness, gallantry,

and incredible good luck." 6 And so it went, a curious
little war inspired by "yellow journalism"* and later
judged the most disgraceful episode in American
journalism. Historians seem to be in agreement in
blaming the newspapers, which are of course ruled by
Gemini.

2. Achievement, Universal Welfare - Pluto in Gemini

Transportation
Communication

Although we certainly have by now a well developed
system of transportation and communication, this was
most distinctly the period that gave birth to most of
our present instruments and forms. There was a virtual
"explosion" in this field which is quite phenomenal in
retrospect. Consider the following:

Bicycles: Although the bicycle had been around for
years, it was said that by 1876 "its time had come".
By 1884 there were 50,000 cyclists, and its popularity
continued to increase. During the 90's, 10,000,000
Americans took to cycling.[7] Quite an increase!

Railroads: As mentioned previously, railroads came
into being during the Taurean Era, but historians say
that the early 1900's were truly the "Golden Age of
Railroads."[8] Popularity and usage increased until
1916,(two years after Pluto left Gemini) and thereafter declined.....very close timing, astrologically.

Automobile: In 1900 there were only 8,000 cars. Within
the next decade, 460,000![10]

Telephone: In 1870 there was no such thing as a telephone. By 1900 we had 19,000 telephone operators.[11]

Consider the following major advances in these areas
when judged chronologically: [12]

*Term coined for misuse of the journalistic power. Al-
though the color blue is usually assigned to Gemini, I
have always believed yellow to be most appropriate as
it relates to "mentality" or"thinking", which is Gemin-
ian. Even school buses are traditionally yellow, so
better fit this designation. Coincidently, Yellow Fever
was a problematic disease during the Pluto in Gemini Era.
14.

Before 1882
C. 1448 - Printing Press
1790 - Steamboat
1826 - Photography
1867 - Typewriter
1876 - Telephone
1876 - Nikolaus Otto constructs prototype of modern
       car engine
1877 - Thomas Edison invents phonograph (perfected 1888)

1882 - 1914: Pluto in Gemini
1883 - Brooklyn Bridge; "Engineering marvel of 19th Cen."
1884 - Eastman patents roll film
1884 - Waterman perfects fountain pen
1885-1887 - Benz and Daimler construct first petrol
       engine autos
1885 - Dictaphone
1887 - Monotype
1887 - First extensive electric streetcar system
1887 - Berliner invents "Gramophone" with disk record
1888 - Motion picture projector
1892 - Diesel patents Diesel engine
1895 - Marconi, wireless telegraphy
1895 - Electric locomotive
1897 - First subway
1898 - "Telegraphone", early tape recorder, by Poulsen
1900 - Photocopy machine
1901 - Radio telephone
1903 - Wright Brothers make first successful airplane
       flight
1906 - Fessenden transmits first modulated radio-wave
       broadcast
1908 - Ford introduces assembly line for Model-T mass
       production
1911 - Automobile self-starter
1913 - Talking motion pictures

After 1914
1922 - Telephoto
1923 - Television
1925 - All electric phonograph
1927 - Transatlantic telephone service
1933 - FM radio
1937 - Helicopter
1958 - Atomic powered submarine
1962 - Satellite Telstar provides direct communication
       between U.S. and Europe

Undoubtedly some have been overlooked, but these are
the major contributions. We have recently made tremen-

dous strides in science, but narrowed to transportation and communication, the Gemini-related areas, recent developments have been largely refinements of the above. Remember too that we are comparing a 450-year period (before 1882, starting with the 1400's) with a 31-year period (Pluto in Gemini), with a 60-year period (after 1914 to the present). Also, some very significant contributions came forth just six years prior to Pluto's entry into Gemini, and so may be attributed to the cusp influence. The odds must certainly be against such a display at the appropriate astrological time, but Pluto always beats the odds!

It is interesting to note too that Uranus, usually credited with the inventive quality, was nowhere on the horizon. During this period it occupied Libra, Scorpio, Sagittarius, Capricorn, and entered Aquarius. The relationship in communication realms just isn't there.

3. Middle Turning Point and/or Crisis - Pluto in Gemini

Midpoint Date: 1897, 1898

The Spanish-American War: 1898

This has been previously discussed under "Annihilation" and is obviously in the "crisis" category. Time-wise we're right on target for the Pluto cycle.

4. Polarity Emphasis - Pluto in Gemini

Polarity Sign: Sagittarius   Ruler: Jupiter

Qualities: Projection, aiming, futuristic, opportunistic, sporty, philosophical, and mental. Its expression is exuberant, optimistic, direct, and ambitious. Keynote: Aspiration

Events: "The Age of Optimism"

Conservation Program Established

Sagittarian enthusiasm is reflected in the names given this period by different authors: "The Progressive Era"13, "The Age of Optimism"14, "The Cocksure Era"15- different writers but similar judgments, and a reflection of the Sagittarian principle emerging.

Also, Sagittarius is sporty, relates to the "great

16.

outdoors" and specifically to wildlife; thus, it was during this period that Theodore Roosevelt established a vast conservation program which promised "protection for all harmless wild things"[16], and huge reserves of land were set aside for National Parks, forests, and wildlife refuges (151,000,000 acres; almost the size of Texas!)

5. Group Activity and/or Movement -Pluto in Gemini

The Edwardian Age
Migration
Immigration
Urbanization
Education Revamped

As the Taurean Period corresponded historically with the Victorian Era, the Pluto in Gemini Period corresponds neatly with the Edwardian Age, dated 1880-1914. The parallel is an interesting one in terms of the Gemini quality. The Edwardian Period is described as a freer era, attempting to shed previous restrictions. The rich and smart were bent on entertaining themselves. They were somewhat shallow, silly, and addicted to senseless games, rushing about spending money. The working class supported this shallowness by vicariously enjoying the lives of the rich through relatives who worked for them. The middle class was described as particularly clever, and this too is a Gemini attribute. Negative Gemini is often shallow, restless, and silly, so the total description is fairly apt.[17]

Since Gemini represents travel and movement, it is not surprising to find the greatest migration and immigration in American History*[18] during this period. The population of Mississippi westward increased from 7,000,000 to over 16,000,000;[19] i.e. it more than doubled within a 30-year span!

Along with this vast movement within the existing population, hordes of newcomers were pouring in. By 1907, well into the Gemini Era, immigration was at an all time peak;[20] and one-third of the total population was of foreign birth.

Another facet of this unbridled movement was urbanization. Historians say that "after 1877 people rushed pell-mell into the complexity of modern civilization."[21]

*"Migration" period: 1870-1890

The socialization of city life is much more appropriate to Gemini, too, than the previously depicted agrarian society. In 1870 only a dozen cities had populations of over 100,000,[22] but as the crowds poured in, new systems of transportation had to be developed to relieve the choked horse-and-buggy streets. By 1888 San Francisco had a smooth trolley system in operation and within two years, 200 other cities had followed suit.[23] Thus, there were mass movements across and into the country, from farm to city, and increased momentum within the cities;-a beautiful example of Gemini nervous energy motivating an entire population toward pure movement.

We also find at this point the development of the department store, the chain store, the shopping center, and the apartment house; all timely for the Gemini quality of "variety".

"Linkage" is another Gemini expression; the basis, in fact, for the communicative and transportation aspects of this sign. Quite suddenly a huge mass of isolated cities and individuals were "linked" through the telephone, railroad, automobile, and new bridges, one of which was the Brooklyn Bridge (1883), considered the "engineering marvel of the 19th Century."[24] The Panama Canal was also constructed during this era; a mammoth undertaking and another example of linkage.

Gemini also relates to teaching, and this was a time of vast change and the beginning of some truly innovative techniques within the school system. Buoyant belief in the power of education sprang up after the Civil War, and by the time a system was established and functioning in all states, it was 1898,[25] the middle of Pluto in Gemini. Pluto obliterates boundaries and it was precisely this which was hoped for. One educator remarked: "The high school shall level the distinction between rich and poor, thus allowing the laborer's son to stand alongside the rich man's son."[26] This is, of course, exactly what happened as Pluto transited Gemini; the intellectual caste system was wiped out as rich and poor alike were taught to read, write, and think.

The new century opened with the usual strict (Taurean) rules: children were to be seen and not heard; to be scrubbed, whipped, and dictated into submission. However, as the decade progressed the rigid code and underlying doctrine of self-reliance became harder to uphold, as well as less necessary. Accordingly, schools were made less oppressive through the addition of pictures

18.

added to drab texts, playgrounds for amusement, and the establishment of groups such as the Boy Scouts and Campfire Girls.

John Dewey founded the first "child-centered" school in Chicago in 1900, completely rejecting the current methods. He firmly believed education must be functional and not simply concerned with rote learning of the Three-R's.

By 1905, another landmark appeared: the first I.Q. tests were developed, hardly surprising under the mentally oriented Gemini influence. Not until Pluto reached another Mercury sign, Virgo, were such revolutionary changes witnessed in our educational system.

6. Non-Correlating Events - Pluto in Gemini

None

7. Sidelights - Pluto in Gemini

* Since the Gemini Third House rules "relatives", we should also note that it was 1905 when Albert Einstein announced his special theory of relativity. Pluto's action is to take us to the heights and depths, and this was certainly a new plateau for the Gemini"relatedness"expression: from neighborhoods to the cosmos in one giant leap!

* Because of the confusions and multiplicities caused by local time zones, in 1883 the railroads instituted Standard Time. An Indianapolis newspaper sarcastically wrote: "The sun is no longer boss of the job. People - 55,000,000 of them - must eat, sleep and work as well as travel by railroad time."[27] Pluto in Gemini, standardizing our lives.

PLUTO IN CANCER (CAPRICORN):   September,1912-July,1939

> "I see one-third of a nation ill-housed, ill-clad, ill-nourished."
>
>   Franklin D. Roosevelt,
>   Second Inaugural Address, 1937

Cancer-Gemini Period: 1912, 1913
Cancer-Leo Period: 1937, 1938
Midpoint: 1925, 1926

Moon-ruled, cardinal-water sign Cancer relates to
hearth and home, women and motherhood, and is the em-
bodiment of family life, security, and tradition. Anything of historical value and patriotic expression is
within the domain of Cancer. It is the essence of nurturing, containing, enclosing, protecting, and turning
inward. Its expression is emotional, delicate, sensi-
tive, romantic, and sentimental.

Major Events and/or Issues

World War I
Isolationism
Wall Street Crash
The Great Depression
Welfare Legislation
Prohibition

1. Annihilation - Pluto in Cancer

World War I: 1914-1918
(Our Involvement: 1917-1918)

Patriotic Fervor
Isolationism

World War I commenced in 1914, the beginning of the
Pluto in Cancer Era. Although the U.S. didn't enter the
fray until 1917 with the sinking of the Lusitania , the
interim was charged with empathetic involvement. As a
matter of fact, the entire war period was fraught with
a patriotic fervor the likes of which we haven't seen
before or since. "Never was a war in history so pas-
sionately embraced in a flash of naive patriotic ec-
stasy as America's war against the Kaiser."[1]. Patriotism
of course, is ruled by Cancer.

It was "America for Americans", as Theodore Roosevelt
put it,[2] and thousands joined patriotic clubs and at-
tended public meetings. Henry Ford instituted a compul-
sory English school among foreign-born employees and
the first thing they learned was "I am a good American".[3]

America was poorly prepared to enter the war in 1917,
but people gave generously of time and money. 23 billion
dollars was raised from a nation whose annual income
totaled less than 70 billion;[4] this surely represents
a strong personal commitment from a great many people.

Wartime laws were passed establishing heavy penalties
for criticizing any aspect of government. A 20-year
20.

prison sentence was possible for advocating reduced production of war necessities.[5] Troublemakers were few, however, as everyone rallied behind the "Grand Old Flag."

While the immediate cause for the war is not as clear cut as in the Taurean and Gemini Periods, a glance through the history books will quickly reveal an un-paralleled expression of devotion to country. Its personality, so to speak, was altogether Cancer-ish.*

So too was the post-war period. The economic boom precipitated by war needs continued, with more and more people partaking of the "good life". Nevertheless, this was described as a period of bewilderment and disen-chantment (moody Cancer). The old ideals of hard work, self-reliance, and faith in God were no longer enough. In contrast to Theodore Roosevelt who led the country through the optimistic and confident Gemini Era, Wood-row Wilson was introspective and complex, quite rep-resentative of the Cancer qualities.

After the war, America in crab-like fashion withdrew into herself in what is considered a period of isola-tion; thus the League of Nations, an organization founded out of concern for a beaten Europe, was defeated

Calvin Coolidge, president in the "Roaring Twenties" summed it all up: "The business of America is business;"[6] self-protective Cancer and business-oriented Capricorn.

2. Achievement, Universal Welfare - Pluto in Cancer

Social Legislation

When Pluto triggered the protective Cancer consciousness the most massive (to date) social legislation in our history was formulated. Security-oriented Cancer is not only protective and nurturing, but can also be quite dependent. These tendencies were bundled into one pack-age through the Social Security Act of 1935. It would be hard to find a neater solution: we were taking care of others and being taken care of at the same time. It was obviously an idea whose time had come.

In addition to the Social Security Bill, we find

*Even the slang expression for our soldiers, "Doughboys", reflects Cancer and its nutritional bond.

umpteen other legislative acts and agencies contrived
to help a hungry, fearful nation; "Mother" taking
care of her brood:

Wealth Tax Act
Banking Act
Public Utility Holding Company Act
CCC - Civilian Conservation Corps
TVA - Tennessee Valley Administration
NRA - National Recovery Act
SEC - Securities Exchange Commission
WPA - Works Products Administration
RFC - Reconstruction Finance Corporation
PWA - Public Works Act
NLRB - National Labor Relations Board
Subsidies to Farmers
FERA, AAA, HOLC, FDIC, CWA

Some critically call this the beginning of a welfare
state[7], but be that as it may, "welfare" is in the
Cancerian domain, whether it be for one's self, one's
family, or one's nation.

This was obviously a period of a mammoth outpouring
of the Cancer quality, and interestingly enough was
brought to a halt in 1938, the end of Pluto in Cancer,
with the Fair Labor Standards Act which marked the end
of the New Deal reform legislation.[8]

Pluto has long since left Cancer, but note that Social
Security is still with us!

3.Middle Turning Point and/or Crisis - Pluto in Cancer
    Midpoint Date: 1925, 1926
    Wall Street Crash: 1929

In October, 1929, the Wall St. Crash (Capricorn rules
"the marketplace" even though Taurus rules "money"
per se) evoked a crisis of world-wide proportions.
Saturn-ruled Capricorn can be indicative of great
suffering, and suffer we did.

The whole world lapsed into bankruptcy and depression,
as world trade declined by 70%. By the early 30's,
American industrial output was ½ that of 1929. Hourly
 wages dropped 60%, white collar salaries 40%, and in
1931 alone, 20,000 persons committed suicide. There
were floods, droughts,and dust-storms. Factories were
idle, and even the flourishing auto manufacturers let
22.

everyone go. Crops went unharvested, milk undelivered; malnutrition, scurvy and pellagra were rampant[9]...all Cancer-related ailments.

With three major planets posited in Cancer, the U.S. was no doubt especially vulnerable in terms of the Pluto transit. Pluto's action is to annihilate and re-establish, and Cancerian problems manifested on a vast scale as well as the solutions which were to follow. Thousands were jobless and homeless. One Chicago official said: "There is not a garbage dump in Chicago which was not diligently haunted by the hungry. Each summer, when the smell was sickening and the flies were thick, there were a hundred people a day coming to the dumps looking for food."[10]

This, then, was the situation which broke up a 50-year romance with businessmen (Capricorn) and led directly to welfare legislation as detailed under "Achievement."

4. Polarity Emphasis - Pluto in Cancer

    Polarity Sign: Capricorn   Ruler: Saturn

    Qualities: Dominance, Control, Limitation, Inhibition, Management, Utilization of Power. Relates to politics and the community at large. Its expression is practical, sensible, and sometimes harsh. Keynote: Hard Work

    Events: The Great Depression
            Strong Government Leadership

The swing to Saturn-ruled Capricorn is easily seen through the privations of the Depression, as previously noted. However, power politics and governmental action in general are also within the Capricorn domain, and the national urgencies of our troubled depression years brought us one of our strongest leaders to date, Franklin D. Roosevelt. He was considered the "Grand Master of Politics",[11] and it was hoped that he could "restore the faith of a nation that seemed almost palsied by fear."[12] Negative Cancer, of course, is exceptionally timid, and in this case, dominating Capricorn was the antidote.

In F.D.R.'s inaugural speech he firmly remarked: "First of all, let me assert my firm belief that the only thing we have to fear is fear itself - nameless, unreasoning, unjustified terror...."[13] It was said that within two weeks the spirit of the country seemed 23.

changed beyond recognition.[14]   Will Rogers quipped that
if FDR had burned down the White House people would
have said that "at least we got a nice fire going"[15]...
Pluto in Cancer-Capricorn, calming a fearful nation
through governmental strength.

Add to this the unprecedented governmental action as
previously noted, and we have a pretty strong case for
the Capricorn expression on a national level.

5. Group Activity and/or Movement - Pluto in Cancer

    19th Amendment - 1920
    18th Amendment - 1919
    Crime Wave

It should be quite clear by now that Pluto was influencing our affairs long before its discovery: we are obviously related to our universe in diverse ways, whether
or not we have the awareness to realize it. In the early
Pluto in Cancer period we hadn't yet discovered our distant friend, but had we known and understood, one could
well have predicted some event of national importance
which would affect women and/or womanhood.

American women had been rather quietly campaigning for
years in an effort to gain the right to vote, and it
seems the mass consciousness finally nodded their way
when strengthened by Pluto's Cancer occupancy. The 19th
Amendment to the Constitution, guaranteeing this right
was passed on cue in 1920.

The 18th Amendment which inaugurated Prohibition is
also on target for our "food and drink" sign though
such a turn of events seems surprising even in retrospect. This amendment was passed in 1919 and repealed
in 1933, so it dominated the national scene for 14
years, a large portion of the Cancer Era. "Moonshining"
and smuggling became big business while hard liquor
became an obsession with many and the consumption of
alcoholic beverages actually increased. No other topic
took up so much space in newspapers and magazines, and
a national slogan was born: "See America Thirst".[16]
Such a monumental concern with "drink" would seem to
qualify for a Cancer-related "group activity" and again,
could only be triggered by Pluto. Neptune, usually associated with hallucinogenic substances, was in Leo
when Prohibition was inaugurated, and in Virgo when repealed....a most unlikely arrangement. Fads and fancies
are Neptune's domain, but when the national law is
24.

involved, check to see if Pluto is on the scene - it usually is.*

This is, by the way, one of the few instances wherein a Plutonian action was not long-lived. I do believe, however, that had it not been also repealed during the Cancer reign, it might still be with us. What is established under Pluto is well established, and we would do well to see that our Plutonian decisions are wise ones.

Since Pluto was discovered in 1930, every horror that occurred during the decade has been placed at Pluto's door: guilt by association. The assignments are slightly misplaced, in my opinion. Pluto represents an unbounded strength, power, and force, but its power is neutral and is simply misused by us unevolved mortals.

The wave of racketeering, for instance, was designated as a Plutonian expression, yet gangsters were not new. What was new was Prohibition, and the gangs simply capitalized on it and grew by leaps and bounds. They organized into alky cookers and rum runners, adulterated genuine liquor, and dominated the speakeasies. By 1925 Al Capone controlled 10,000 speakeasies in Chicago and had an "army" of 700 men. Rival gangs, named for their leaders emerged: the Aiellos, the Gennas, the O'Banions.[17] Now isn't this phenomenon simply a variation on the old Cancer "family" theme? Obviously Pluto was in effect during earlier (and later) periods, and there was no particular emphasis on crime through organized gangs until Prohibition came along. Let's place the blame where it rightfully belongs...with Pluto in Cancer. Each planet, each sign, has its negative potential, and this is simply an example of negative Cancer, energized by Pluto.

One last charge which we should also clear up is the usual assignment to Pluto as the "kidnapper" since again, there was a wave of kidnapping during the 30's, shortly after Pluto was discovered. There is some grain of truth here, but once more it has been slightly distorted Each planet can represent some lawless activity: Mercury, the "con" man, Mars, the brute, etc. Well Pluto, acting negatively can be the thief, and what is kidnapping after all, but stealing from the parents,Cancer-

*We are currently (1973) witnessing many Neptune-inspired laws due to its Sagittarian transit, but ordinarily this is not the case. Many of these decisions will undoubtedly be repealed.

Capricorn, "Mother-Father"? Hairsplitting? Maybe..but were we to assign "kidnapping" to the Pluto function alone then we should have the problem to contend with during all the transits, and Pluto should be involved in each case, and the theory simply doesn't hold. "Thievery" is more exactly correct, and as will be seen, this assignment does carry through.

6. Non-Correlating Events - Pluto in Cancer
None

PLUTO IN LEO (AQUARIUS):   October, 1937 - July, 1958

> "I should like to have it said of my first adminis-
> tration that in it the forces of selfishness and of
> lust for power met their match. I should like to
> have it said of my second administration that in it
> these forces met their master."
>
>   Franklin D. Roosevelt[1].

Leo-Cancer Period: 1937, 1938
Leo-Virgo Period: 1956, 1957
Midpoint: 1947, 1948

Sun-ruled, fixed-fire sign Leo relates to authority, power, royalty, and recreation. Its essence is vital- izing, superiorizing, centralizing, and vivifying. Its expression is energetic, colorful, bombastic, dominating, extroverted, and flamboyant.

Major Events and/or Issues

World War II
Atomic Energy
Scientific Advances

1. Annihilation - Pluto in Leo

World War II: 1939-1945
(Our involvement: 1941-1945)

Struggle for World Power
The Atom Bomb

Like a fighter coming out when the bell rings, the on- slaught began shortly after Pluto's entry into Leo: Germany marched into Poland, Italy into Albania, and

Russia into Poland....the power struggle had begun.
Hitler's credo, "Today we own Germany, tomorrow the
whole world!"[2] is the epitome of negative Leo, lusting
for power.

The fear of total domination was a real one, and even
Bertrand Russell, a life-long pacifist, remarked: "Lib-
erty cannot be preserved without military struggle, and
will die out unless totalitarianism is defeated."[3]
American sentiments were much the same, and in 1940
Walter Lippmann wrote: "Our duty is to begin acting at
once on the basic assumption that before the snow flies
we may stand alone and isolated, the last great Democ-
racy on earth."[4]

Germany quickly gobbled up Austria, Czechoslovakia, Poland, Denmark, Norway, and Sweden. On the eve of the in-
vasion of France Hitler spoke: "Soldiers of the West
Front! The hour for you has now come. The fight beginning today decides the fate of the German nation for
the next 1,000 years."[5]

Thus did the Germany-Italy-Japan Axis seem hell-bent to
conquer the world, while the remainder declared a fight
to keep it free, a goal beautifully idealized by Franklin Roosevelt's State of the Union message in 1941: "In
the future days, which we seek to make secure, we look
forward to a world founded upon four essential freedoms.
The first is freedom of speech and expression - everywhere in the world. The second is freedom of every person to worship God in his own way - everywhere in the
world. The third is freedom from want....everywhere in
the world. The fourth is freedom from fear....anywhere
in the world.....To that high concept there can be no
end save victory."[6] And so we find the battle lines
clearly drawn between Aquarian and negative Leo principles; opposition on a grand scale, triggered once again
by extremist Pluto.*

At war's end in Europe, atrocities unprededented in
human history were revealed in the German concentration
camps. The calculated extermination of 6,000,000 Jewish
people, the inhuman treatment of prisoners, medical
experiments done for the benefit of "The Master Race"..
these were termed "crimes against humanity",[7] and hu-

---

* Slang expression for American soldiers in this war
was "G.I." for Government Issue....another echo of
our heightened authority consciousness.

manity", of course, is the very core of the Aquarian principle.

The Atom Bomb was dropped on Hiroshima in August, 1945, and brought the war to a conclusion. Although this was the end of the war, it marked a special point in time as we thus began the Atomic Age. This Plutonian event was the opening statement in the chapter on Atomic Energy.

The Korean War: 1950-1953

This is the only Pluto period with two major wars, and perhaps this too is significant. Any Leo emphasis is always full to bursting with energy; the results are fruitful even if, as in this case, the fruits are the bitter ones of dissension.

The Korean War, however, occurred in the latter half of this era and thus is expressed more distinctly through the Aquarian principle. This war, or "conflict" as it was commonly referred to (and may thus even by label be more Aquarian), was supported by the United Nations, with each nation sending a percentage of troops. Even in warfare, obviously, the Aquarian expression of co-operative effort prevails.

The war ended without victory and through joint agreement to divide the country at the 39th parallel; a rather unemotional note on which to end a war, but perfect for the Aquarian reference.

2. Achievement, Universal Welfare - Pluto in Leo

> "Sixteen hours ago an American airplane dropped one bomb on Hiroshima...It is a harnessing of the basic power of the universe. The force from which the sun draws its power has been loosed against those who brought war to the Far East."
>
> Harry S. Truman

Atom Bomb - Beginning of Atomic Age

Since Pluto is a planet of gigantic power and tremendous force, there is an unquestionable link with Atomic Energy. The shape of an atomic explosion, in fact, brings to mind the mushroom which is ruled by Pluto.

28.

However, there is yet another factor involved which is particularly pertinent to astrologers: the timing of this event, which was at the approximate midpoint of the Pluto in Leo-Aquarius Era.

Leo is ruled by the Sun, the "center of power". Its symbol, in fact, looks exactly the same as a graphic representation of a simple atom:

Sun Symbol   Representation of H-Atom

Also, the Leo polarity sign, Aquarius, is especially noted for scientific prowess, so it would appear that we are covered on all fronts for such an event to have occurred during this Leo Period of "greatness".

3. Middle Turning Point and/or Crisis - Pluto in Leo

Midpoint Date: 1947,1948

World Leaders Replaced - 1945
Atom Bomb Dropped - 1945
United Nations Founded - 1945

There is no disputing the fact that 1945 was an important year, if not the year of the century. This is approximately the middle of the Pluto in Leo cycle, and also approximately the middle of the century, so this may have some bearing on the case. Consider the following mind-boggling events in a short 12 month period:

World Leaders Replaced: At the Leo-Aquarius turning point, all the dominating world figures toppled from power at once, one way or the other: Roosevelt died, Mussolini was murdered, Hitler died (suicide?), Churchill was voted out of office, Hirohito was defeated. Only Stalin remained (explanation under "Polarity Emphasis").

Atom Bomb Dropped; War Ends   This is an extremely Plutonian event from any standpoint, as previously noted. It was said of Hiroshima that "every living thing was dead or waiting to die" and that

29.

"A city died giving birth to an era."[8] One is reminded here of the Pluto-sign Scorpio symbol: a phoenix arising from the ashes...life from death.

United Nations   Here is a fine example of the posi-
Founded: tive Aquarian principle in action;
                     a universal cooperative effort to
unite the world. It is interesting to note that Wilson's
attempts in this direction through the League of Nations
though laudable, just didn't come off. Timing, the crux
of Astrology, is very important as witness the results
of the same goal pursued at different times.

4. Polarity Emphasis - Pluto in Leo

> "Our strength lies, not alone in our proving grounds
> and our stockpiles, but in our ideals, our goals,
> and their universal appeal to all men who are
> struggling to breathe free."
>
>   Adlai Stevensong

Polarity Sign: Aquarius   Ruler: Uranus

Qualities: "Cooperation of many units as one", individ-
             uality, pulling-together. Relates to clubs,
             groups, friendship, and humanitarian en-
             deavors in general.Its expression is scien-
             tific, understanding, detached, and inde-
             pendent. Keynote: Cooperation

Events: Scientific Advances
             Communistic and Freedom Principles Espoused

Aquarius rules science, and after 1945 there was a tremendous growth in scientific accomplishments. Some of
the significant advances: [10]

                Atom Bomb, H-Bomb
                TV became commonplace
                Sputnik (1957)
                U.S. Satellite (1958)
                Telstar, Tiros, Marmer, Echo (satellites)
                Laser
                Heart surgery
                Electronic Computers
                Man-made giant molecules of RNA
                Anti-polio vaccine
                Atlantic missile test center inaugurated
                Micro-electrics developed

30.

Notice here that Aquarius encompasses <u>all</u> of science, while the Gemini Era gave us scientific advances in transportation and communication, primarily, while the Virgo-Pisces Era, next to come, seemed to zero in on computers and medicine. The Aquarian scientific turn-in-the-road comes through loud and clear with the statement by historians that "science transforms life."[11]

Aquarius, being group-oriented, also is supportive of the communistic principle, and being Uranian, supports individual freedom; a seeming contradiction, yet it does hold true. Aquarius rules all of humanity and at the same time, the absolutely unique individual; the "all for one and one for all" expression. After 1945 there was a mass movement toward these two ideals. Communistic and free countries flourished while dictator ships and monarchies were abolished.[12]

<u>India</u>: Gained freedom in 1947. Divided into two independent nations: Dominion of India and Moslem State of Pakistan.

<u>S. America</u>: End of all but three dictatorships; replaced by popular democratic regimes intent on economic and social advance.

<u>Germany</u>: Soared to economic leadership in Europe with system based on free enterprise.

<u>Israel</u>: Established as a free state.

<u>China</u>: Became Communist. All landed people executed for collectivization of land. 800,000 to 20,000,000 executions of landowners. 450,000,000 peasants herded into communes and cooperatives. Children given to state to free mothers for work in fields.

<u>England</u>: Became Socialist with overthrow of Churchill.

<u>Canada</u>: Coming of age symbolized by appointment of first Canadian as Governor General instead of royalty or a British Peer.

<u>Japan</u>: Hirohito's son repudiated doctrine of divinity after the war.

<u>Indonesia</u>: First free elections; a "Guided Democracy".

<u>Indo-China</u>: Free of French rule; split into four parts: independent Laos and Cambodia, North and South Vietnam.

<u>Burma</u>: Gained independence in 1948.

<u>Malaya</u>: Gained independence.

<u>Singapore</u>: Gained independence.

<u>Africa</u>: In 1950 only one-tenth free. By 1961 over two-thirds.

<u>United States</u>: 22nd Amendment to Constitution limits presidency to two terms.(1951)

Astrologically, the emergence of the Aquarian principle is clearly evident at this point in time. Some believe it the beginning of the Aquarian Age, others that it denotes the Aquarian Epoch of the Piscean Age, and for those who work with polarity signs, the Pluto in Leo-Aquarius presents still another possibility. The reader may take his choice, but note this: trying to limit this historical survey to the United States alone was simply impossible during this era. When the Aquarian expression was energized, for whatever reason, significant events were of universal importance and could only rarely be viewed within the national context.

5. Group Activity and/or Movement - Pluto in Leo

   Totalitarianism

The previous Pluto eras show some display of the two involved signs throughout the period, but in this particular case the expression is much more distinctly divided. Therefore, to observe a strong Leo manifestation, we must regress to early Pluto in Leo-Aquarius.

The existing world situation in 1937 was typically Leo power-oriented. Totalitarianism flourished and the tight authority of monarchs or dictators dominated the entire earth.

Even in the United States, one of the few democratic regimes, we gave our president unprecedented powers and ultimately elected him for four terms, unheard of before or since. Also, due to wartime needs after 1941, the entire nation cooperated with the most stringent rules ever handed down by the Federal Government:[13]

>   Food was rationed
>   A curfew was enforced
>   Pleasure driving was banned

People were "frozen"on essential jobs
The military draft was inaugurated
The government seized Montgomery Ward
Japanese-Americans were put in intern-
       ment camps
Certain clothing items were rationed

In other words, whether free or not, whether forced or
self-enforced, circumstances brought into being a
strong authority consciousness: extremist Pluto pushing
to the limits once more, and this time in "kingly" Leo.

6. Non-Correlating Events - Pluto in Leo

None

PLUTO IN VIRGO (PISCES): November, 1956 - August, 1972

> "Ask not what your country can do for you - ask what
> you can do for your country."
>
>   John F. Kennedy[1]

Mercury-ruled, mutable-earth sign Virgo relates to the
common man, jobs, health, pets, hygiene, service, healing, and all mental work involving much detail. It is
the essence of "take-it-apartness", analyzing, separating, "subsidiary-ness", subservience, serving, tearing-
down. Its expression is dutiful, orderly, critical, and
mental. Keynote: Detail.

Major Events and/or Issues:

Vietnam War
Computerization
Drug Problems
Ecology Movement
Moon Landing

1. Annihilation - Pluto in Virgo

The Vietnam War: 1962 (?) - 1973

Advisors
Commitment
Protest Groups
P.O.W.'s

The vague beginnings of this war immediately proclaim
the Pisces-Neptune influence at hand. Accustomed to
fighting wars which began with a date and a cause, in
this case, an ultimately disillusioned American public
didn't ever quite know when or why they had started
fighting; a perfect picture of Neptunian confusion.

Its origin, as we can now trace it, was deceptively in-
nocent: Truman sent 35 military advisors to Vietnam,
a reasonably small contribution toward helping this
small struggling country. Eisenhower later upped this
number to 500, and under Kennedy, our "advisors" totaled
16,000. By 1966, during Johnson's administration, we
had involved ourselves to the tune of 357,000 men.[2]
By this time America was not only "advising", but had
completely enmeshed herself in the war, and negative
Virgo is famous for precisely these tendencies: giving
advice and poking her nose into other people's business.
The trap had been set, and we stepped right into it.

This unpopular war was marked by protest, counter-pro-
test, and anti-protest-protest.[3] Significantly, some of
the most adamant leaders of the anti-war movement came
from the Piscean ranks; among them: Dr. Benjamin Spock,
and Priests Daniel and Phillip Berrigan. The government
countered their objections with declarations of "commitment", and so, as with Leo-Aquarius, the lines were
drawn: the duty-bound Virgo consciousness vs. compas-
sionate Pisces.

It was called an "immoral"[4] and "nasty little war", and
the nation divided into "doves" and "hawks".[5] (Virgo
rules birds too.) Flags were torn at demonstrations and
many young men burned draft cards and simply refused to
fight. Contrast this situation with light-hearted
Gemini's "Splendid Little War", powerful Leo's fight
over world domination, or the patriotic Cancerian fray
with the flags waving proudly.

Another important facet of this war which seemed peculiar to it alone, was the concern with the P.O.W.
Pisces re-enters the picture here, since "incarceration"
is a Piscean 12th House condition. The welfare and treatment of these men became a national preoccupation, and
bracelets with their names were sold as constant re-
minders of their imprisonment. When finally released
they were ushered home with great fanfare, and these
men became, for the first time in any war, our heroes.

Pluto moved out of Virgo permanently in August, 1972,
and the war ended five months later; rather close

34.

timing, astrologically.

2. Achievement, Universal Welfare - Pluto in Virgo

Medical Advances
Computerization

While medical achievements are nothing new, it does
seem that the 60's were unusually productive. Some of
the significant advances:

The Salk polio vaccine was developed in 1955, one year
early for Pluto in Virgo, but polio was considered
conquered only after 1967,[6] well into the Virgo Era.
A disease which had crippled untold thousands was virtually wiped out.

After 1955, tranquillizers became common[7] thus revolutionizing care for mental patients and allowing many to
carry on with everyday life by minimizing stressful
reactions.

Heart and kidney transplants were performed.[8] While
these revolutionary procedures did not always guarantee
lasting success, the results were promising. Use of
pace-makers began, prolonging life for some who would
otherwise not have survived. Confronted with these medical possibilities, we were then forced to settle some
serious moral and ethical questions: when is the moment of "death", and when, as in the case of abortion
(a later issue) is the moment of "birth"? Pluto in
Virgo made health and morality national issues.

The"S.S.Hope", a hospital ship, was sent around the
world offering medical aid and education for needy,
isolated people; a beautiful expression of the Virgo-
Pisces affinity for concerned healing.

Another gigantic effort in the direction of Universal
Welfare was the establishment of the Medicare program;
in Pluto terms, "health care for the masses."

The 60's were also hailed as "The Computer Age", and
we're right in tune again with our mathematical, method-
ical, orderly Virgo influence. Being thus freed from
many laborious record-keeping tasks,we must surely
reeognize such a development as a giant step forward,
although there were some negative ramifications. People
felt dehumanized by this creeping computer technology,
and Eric Fromm remarked: "A specter is stalking in our

midst...a completely mechanized society...man is being
transformed into a part of the total machine."9 This
sort of social transformation on a vast scale very accurately reflects extremist Pluto working through precise, exacting Virgo.

While Virgo is not usually considered in terms of social
consciousness, laws enacted for the benefit of the common man do suit Virgo-Pisces, and we find a flood of social legislation in this era. In 1960 there were only 45
domestic social programs; by 1969, 435. Many of these
programs related specifically to Virgo, and must be acknowledged as a form of "achievement": Medicare, massive
aid to education, programs to promote job training, programs to combat pollution, etc.

3. Middle Turning Point and/or Crisis - Pluto in Virgo

Midpoint Date: 1964

The Vietnam War (previously discussed)

4. Polarity Emphasis - Pluto in Virgo

> "Some see things as they are and say why. I dream
> things that never were and say, why not."
>
> Robert F. Kennedy[10].

Polarity Sign: Pisces   Ruler: Neptune

Qualities: Mutable-water sign Pisces is the sign of
              limitation and confinement, including hos-
              pitals and prisons. It rules mysticism and
              the psychic realms, and is the essence of
              universality, non-separateness, and a "re-
              ducing-to-one-common-level". Its expression
              is sensitive, intuitive, sympathetic, and
              compassionate. Keynote: Solitude

Events: Psychic Awareness
              Civil Rights Movement

Disillusioned by the destructive forces that seemed to
accompany our burgeoning technological society, many
turned away from our materialistic, "rational" beliefs.
Some "tuned out" with drugs, and dangerous as it was,
doors were opened on new frontiers of the mind. The
psychic barriers, for good or ill, had been broken.

However, with or without drugs, a new fascination with the mystical experience emerged. The "irrational" became popular, and some experts view investigation into psychic areas as the most important advance in recent times. According to Time Magazine one scientist remarked: "It may be that aspects of mysticism totally outside science may come back and be incorporated within its framework."[11]

In light of our materialistic, scientific society, these are surely revolutionary stepping-stones, and unquestionably Piscean in nature.The barrier-breaking action of Pluto is in evidence once more.

Another significant movement during the 60's was the Civil Rights Movement, but again (the Neptunian influence in this period?) it is difficult to pin it down. Is it best described as an "achievement", a "movement", or a demonstration of the Piscean polarity influence; an expression, perhaps, of "non-separateness?"

To make matters even more complicated, the Alabama bus boycott led by Dr. Martin Luther King Jr. in 1956 marked its beginnings as a national movement. This was the same year that Pluto was moving from Leo-Aquarius to Virgo-Pisces. Thus we find that the humanitarian overtones of this movement can be well placed in the Aquarian context, while the quest for equality (reducing-to-one-common-level) is timely for the Piscean emphasis in the 60's.

There are obviously many influences and effects converging here and the astrological assignments are not clearly defined for the Pluto reference. It is perhaps best understood just as it stands; in the midst of many and diverse configurations.

5. Group Activity and/or Movement - Pluto in Virgo

    Drug Addiction
    Ecology Movement
    Educational Changes

Virgo rules drugs, and Pisces specifically rules "dope", so either way we find a connection with drug addiction which was surely the major domestic-social problem of the 60's.[12] Virgo also relates to health, of course, and Pluto puts it on a grand scale, so the component parts of the situation piece together only too neatly.

LSD, marijuana, "speed", heroin, etc. all became part and parcel of the public scene. What was once thought of as a rare and tragic habit became an all too common problem. Because the drug scene was well timed to the Plutonian influence, we must expect this to remain with us for some time to come. What is established through the Pluto influence becomes a way of life.

Another national preoccupation was the ecological move-ment, originating with Rachel Carson's book <u>Silent Spring</u> (1962) in which she warned of the dire conse-quences of the indiscriminate use of insecticides.[13] As of that time, the march was on, and we were admon-ished in true Virgo fashion that we had wasted our re-sources, dirtied our world, and it was time to clean up the whole mess. Virgo says "let's clean up", but <u>Pluto in Virgo</u> says "let's clean up the <u>world</u>."

Daily announcements of the air pollution index became a routine part of the weather forecast, and abolition of pollution itself, in all forms, became a prime goal for large groups of people. As so often happens with Pluto, we developed an "obsession": "Don't litter", "don't burn leaves", "don't pollute the water", etc. In fact, the electorate made garbage one of the main issues in the New York election of 1969.

Concurrent with the negative admonitions was a whole movement toward the "natural" and healthy life. Organic gardening became popular, health foods became common-place, great masses took to camping (and some to the hills for good), and we finally began to appreciate the fact that many animal species were in grave danger of extinction.*

"War" was declared on cigarette smoking: ads were ban-ished from T.V. and the Surgeon General entered our lives in the form of a warning on every pack of cigar-ettes: "Smoking is dangerous to your health".

This was the decade, too, when dieting became some sort of national pastime. "Weight Watchers" had a timely be-ginning in 1963 and in a few short years became a na-tional institution. Numerous other diet programs swept the country and we had The Air Force Diet, The Drink-ing Man's Diet, The Stillman Diet, etc. Remember when

*Any emphasis in Virgo or Sagittarius seems to affect animals in general, though the latter is more strictly applicable to "wildlife"; the dividing line is rarely distinct.
38.

people used to gather for a sociable repast and no one was dieting? That was _pre_ Pluto in Virgo.

Because Virgo is Mercury-ruled, the presence of Pluto in its sign brought transformation in many Mercury-ruled institutions; most notably, our schools. The fact that these changes were frequently accompanied by strife and upheaval must be credited to Uranus, which also passed through Virgo in the 60's.

We had teacher's strikes, contention and violence on the college campus, and desegregation in the public schools; the latter, a situation which proved in too many cases to be a volcanic one.

Now that the smoke has cleared somewhat, we can observe new sprouts from what appeared to be violent seeds: students now actively participate in their own educational interests, many racial and neighborhood boundaries have been obliterated, we have "schools without walls" and "open classrooms". When Pluto dominates, the barriers go, and with the help of Uranus, they went very quickly.

Of course the Post Office is ruled by Mercury also, and it _was_ re-vamped (zip codes are a case in point here), yet certainly without the devastating experiences which beset the schools. Just how one makes this fine distinction in advance is still open to debate.

6. Non-Correlating Events - Pluto in Virgo

Moon Landing

Perhaps the most significant event of the 20th Century was landing men on the Moon, July 20, 1969. I can see no astrological correlation for this event, either through Pluto or any other _single_ configuration at the time (Uranus in Libra just doesn't fill the bill.)

One explanation offered is that this related to the beginning of the Aquarian Age, possibly dated February 4, 1962 with an alignment of seven planets in Aquarius.

Another possibility is that the explanation will be found outside the present range of astrological knowledge; specifically, when Vulcan, new ruler of Virgo and ruler of the atmosphere[14] is discovered. It is relatively easy to detect Pluto's handiwork in retrospect, yet before its discovery, these events must have been mis-labeled by astrologers who tried to fit them into a limited frame of reference. With the Virgo ruler in limbo we can be assured that there are still pieces missing from our Cosmic Puzzle.

7. Sidelights

*In terms of the Virgo influence on warfare, it is interesting to note that in 1971 we were in the process of destroying a stockpile of biological weapons; in other words, a means of "germ warfare". This was two years after President Nixon renounced all such weaponry. The significance here, of course, is that it was considered <u>at this time.</u> Lest we think that it was a less than serious project, the cost of simply destroying these deadly organisms was <u>$11,000,000</u>. This is a Virgo expression in its most destructive form, and a better example of misdirected energy just couldn't be found. This was the time to attack disease, not create it! With Pluto in Virgo we could have had undreamed of success.

* When we find examples of anything "super", "fantastic", etc., we also most usually find a strong Pluto link. <u>Time</u> magazine labeled Secretariat "Super Horse" and sure enough, he was born in March, 1970 [15], the Pluto in Virgo period, and tagged with a singularly Virgoan name. (His popularity, of course, rests with Neptune in Sagittarius, which should idealize horses and related matters.)

PLUTO IN LIBRA (ARIES):   October, 1971 -   1984

> "Peace and justice are two sides of the same coin."
>
> Dwight D. Eisenhower[1]

Libra-Virgo Period: 1971
Midpoint: 1977, 1978

Venus-ruled, cardinal-air sign Libra relates to the fine arts, marriage, partnerships, legal matters, and other people in general. It is the essence of beauty, harmony, balancing, adjustment, evaluation, comparison, and reciprocalness. Its expression is pleasant, indecisive, fair and agreeable. Keynote: Harmony

40.

1. Annihilation - Pluto in Libra

Even though Libra is traditionally a peace-loving
and harmonious sign, there are a few things which will
provoke wrath; among them,"peace" and also "justice".
The negative Libran consciousness will demand peace if
it must kill to achieve it! This is, of course, the
Pluto in Libra snare, ready and waiting for an un-
witting victim.

We must beware of haggling over peace treaties to the
point of violence; the confrontation at Wounded Knee
is a good recent example (even though the peace treaty
in question was 100 years old!). The bombing in Cambodia
is another case in point, since our leaders justify
this action in terms of peace treaty violations. Unfortunately, this is _precisely_ the sort of situation
which could readily erupt into another all-out war during this time when we are particularly vulnerable to
Libran negativities.

Diplomatic blunders and legal misunderstandings are also
good timber for a Libran fire, and we should be especially wary of over-reacting in these instances.

With Pluto in Libra we can _fight_ about peace or _bring_
about peace. The reins of power have been temporarily
but firmly placed in our hands.

2. Achievement, Universal Welfare - Pluto in Libra

Mrs. Lyndon Johnson's campaign to "beautify America"
and the formation of the Peace Corps were projects
initiated during the 60's, our Pluto in Virgo Period,
and both efforts have sadly faltered. The motivation
was excellent, but the timing poor, astrologically
speaking. Pluto in Libra would support such movements
beautifully should anyone decide to revive the notion,
and let's hope someone does.

Cooperation is one of Libra's strongest expressions,
and Nixon's trip to China in '71 was an opening chapter
in this regard. We should witness much negotiating and
diplomatic interaction during this period, along with
the possibility of totally new dimensions in inter-
national relationships.

While the United Nations is already well established,
its jurisdictional functions could well be re-vamped

or enhanced at this time, thus bringing forth a power-wielding World Court above and beyond the current structure.

In the same vein, our own national judicial system is also due for some timely changes. With Neptune in Sagittarius, lending us idealism in terms of the "spirit" of the law, and Pluto in Libra influencing us toward change in the actual process of litigation, we can hardly avoid some unique accomplishment in this area. "Justice" as an ideal and as a goal will monopolize the mass consciousness.

Libra is the diplomat par excellence, so we could well achieve some new plateau in the "art of getting along with others". There should be some new scientific know-how in this regard, and people working in these areas (marriage included) are in line for a good deal of support, recognition and revolutionary changes. Watch for similar revolutionary changes in the fine arts in general.

In terms of the polarity sign, Aries, we can expect much progress in treatment and/or prevention of the aging process, since Aries represents "eternal youth". Look for new techniques here, along with new attitudes.

Other than Aquarius, there is no sign more oriented toward individual freedom than Aries, and toward the latter part of this period we could well witness some legislation in the interests of individual rights. The mass consciousness will lean heavily toward the ideal (though in the Pluto reference it becomes a goal) that we are a "free and independent people". Negative Libra can be quite manipulative, and when and if this conditon becomes apparent on the national scene, through government or circumstances in general, the nation will react through the Aries principle.

3. Middle Turning Point and/or Crisis - Pluto in Libra

Midpoint Date: 1977,1978

Keeping in mind the fact that in every Pluto Era observed to date we can readily detect a distinct turning point toward the polarity sign, we can anticipate a special Arian emphasis to emerge during the latter portion of this period.

Keeping in mind also that the midpoint date does not always exactly coincide with the "crisis", we can't be precisely sure of the timing, but a likely choice might be '75, when Uranus moves into Scorpio thus bringing a new power potential into the picture.

As to the exact nature of this crisis, and staying within the Libra-Aries framework, there would seem to be two possible choices: (1) A conflict, according to the possibilities as previously discussed; (2) An energy crisis of rather dynamic proportions. Since Aries rules energy, per se, and we will be heading into an Arian period, this would be the most likely eventuality. Also, it seems rather appropriate that as we near the end of the Piscean Age, we are running out of liquid fuel. Research and development of new energy sources are clearly in order. Returning to older and simpler life-styles is simply not feasible, at least on astrological grounds. We must always move toward the new experience, and the directional indications impel us toward the Aquarian-related energy sources.

4. Polarity Emphasis - Pluto in Libra

Polarity Sign: Aries   Ruler: Mars

Qualities: Sign of initiative, aggression, and pure activity. Relates to the military and leadership in general. It is the essence of impulsive action, pioneering, "get-there-first-ness", and "out-rushing-ness." Its expression is quick, alert, brusque, independent, and competitive. Keynote: Leadership

In previous Pluto Eras, a certain "testing" or challenge of the polarity sign can be noted: for Scorpio to "get rid of waste" (slavery), for Capricorn to limit and manage (the New Deal surfaced), for Aquarius to defend humanity (World War II), etc. It is final examination time, and the Pluto tests are tough. This martialing of energy, however, with which we meet the task at hand ultimately builds new muscle and we are the stronger for it.

The current period will test our ability to mobilize our energies very quickly to meet any national emergency. The Aries expression is first and foremost "quick action", so this we must be prepared to do,

43.

but not, hopefully, with the Aries tendency to rush about and act on the first solution which comes to mind.

Our pioneering spirit will also be challenged, and we will undoubtedly be caught up in investigations of heretofore unthought of matters. When detecting, investigative Pluto energizes our pioneering sign, Aries, something new is sure to come of it. Speculation as to exactly what this might be is difficult, since it is undoubtedly beyond the range of common knowledge at this time.

Finally, our leadership abilities and national leaders in particular will find themselves tested. The Watergate investigation comes to mind here, quite aside from the Libran legal implications. It would seem on the surface to be a problem facing the White House and President Nixon in particular, having strong political ramifications, and so it is; yet the astrological implications are much broader.

When we test Aries, we are testing leadership in general, and that includes all leaders on all levels. It simply becomes more obvious and more important on the national level. Anyone assuming a position of leadership during this period will have to prove his abilities beyond the shadow of a doubt, and those who survive may prove to be quite exceptional in this regard. New standards for those who aspire to lead will undoubtedly emerge. Just "anyone" won't do with Pluto at the helm.

5. Group Activity and/or Movement - Pluto in Libra

Pluto represents anything "fantastic" or "spectacular", and its position in Libra brings forth litigation, so we must quickly return to Watergate and the situation of "fantastic litigation". Daily exposure of the Watergate hearings provided the American public with a "crash course" in law as we were (and continue to be) bombarded with legal proceedings and terminology. Where Pluto in Virgo brought the Surgeon General into our lives, Pluto in Libra has made the Attorney General a household acquaintance . The fact that the majority of men involved in Watergate are lawyers and advertising men strikes still another Libra-Aries chord, and as long as Pluto tenants Libra, both these professions are due for some penetrating scrutiny.

Marriage is also within the Libran domain, and before

Pluto moves to the next sign, the institution as we know it will have changed beyond recognition. Pluto in Virgo brought us "open classrooms," and Pluto in Libra has already brought us "open marriage", but this is only the beginning. Pluto decrees that the barriers must go, so "marriage without bonds" should be the new expression. Both marriage and divorce laws will be somewhat obliterated, and whatever shreds are left will be standardized on a national level.

Most or all of the Venus and Mars occupations will be drastically altered: actors, beauticians, diplomats, mechanics, doctors, dentists, butchers, the military, etc. Note here that already we have witnessed the abolition of the military draft, acupuncture emerging as an alternative to usual medical techniques, and the meat industry challenged by a national boycott.

6. Non-Correlating Events - Pluto in Libra

None to date

7. Sidelights - Pluto in Libra

* A slight digression, but perhaps important, is the apparent Pluto relationship to our national leaders who are obviously linked to any influence which functions on a national level.

It has been observed that every 20 years when Saturn and Jupiter conjunct in an earth sign, the president elected for that term of office dies. while this has been a fairly regular event, the conjunction doesn't always coincide with the exact year of election or year of death, though the relationship is close.

It occurred to me that the 20-year cycle theory may have evolved since it is approximate to Pluto's cycle, and that this may be a further manifestation of the Pluto annihilative quality. It is submitted here simply as a theory which will bear watching.

Notice that not one Pluto period was skipped. Also, in the two cases with two presidents each, one of each pair had an extremely short term. Harrison died after only one month in office, Garfield after only six and one-half months.

President	Year of Death In Office	Year Elected Preceding Death	Saturn-Jupiter Conjunction	Pluto Sign Year Death Occurred
Harrison	1841	1840		Aries
Taylor	1850	1848		Aries
Lincoln*	1865	1864	1861 (Virg.)	Taurus
Garfield*	1881	1880	1881 (Taur.)	Taurus
McKinley*	1901	1900	1901 (Cap.)	Gemini
Harding	1923	1920	1920 (Virg.)	Cancer
Roosevelt	1945	1944**	1940 (Taur.)	Leo
Kennedy*	1963	1960	1960 (Cap.)	Virgo

* Assassinated   ** Also in 1940

It is also an interesting coincidence that in some of
these cases, the planetary ruler of the Pluto sign was
also in evidence: Lincoln was shot in a theater (Venus),
Garfield in a railway station (Mercury: 1881 was cusp
of Gemini), McKinley while on a trip to the Buffalo
Exposition (Mercury), and Kennedy's assassin hid in a
bookery (Mercury).

PLUTO IN SCORPIO (TAURUS):   December, 1983-January,1995

> "Everything science has taught me - and continues
> to teach me - strengthens my belief in the contin-
> uity of our spiritual existence after death.
> Nothing disappears without a trace."
>
> Wernher Von Braun[1].

Pluto-ruled, fixed-water sign Scorpio relates to the
occult, death and matters connected with, sex, creativity, certain branches of science, detecting, and
in general all things which are hidden or private. It
is the essence of "throwing-out-ness", elimination,
"shedding-the-old", and "getting-rid-of-uselessness".
Its expression is forceful, determined, passionate,
and powerful. Keynote: Strength

In Scorpio, Pluto has reached its own sign, so the
ramifications of this transit are likely to be very
significant. We can now consider some future possibilities in terms of previous patterns, keeping in mind the
fact that these conjectures have not been unduly scru-
tinized with regard to other configurations.

Expect some or all of the following to become national
issues, concerns, or realities:

Conflict or destructive activities which emerge due to negative expression of the Scorpio principle.

Transformation in the Scorpio-Taurus occupations: life insurance salesmen, funeral directors, miners, occultists, researchers, farmers, bankers, florists, etc.

Governmental domination, especially through the Secret Service. With Neptune in Capricorn at this time, be aware of the possibility of "glorifying" government so much that we allow too much power to be placed in unworthy hands. Realism is called for along with an awareness of the Scorpionic tendency to "invisible" government.

The Military and/or other militant organizations become more active and possibly reorganized or transformed.

The Supernatural becomes supremely natural; spiritualism and occultism become part and parcel of our daily life.

Death becomes part of the mass consciousness and is elevated to a new plateau in understanding. The "right to die", euthanasia, and national life insurance may all be public issues, and more than likely the boundary between the dimensions we term "life" and "death" will be obliterated.

Natural Resources will be in great demand; we will need to produce, make secure, and demonstrate an ability to harness our industrial forces. Building and construction should be at an all-time high.

Money, credit, banking, and our system of taxation will be challenged and probably revolutionized.

Achievements are possible in investigative and research realms, or any field which unearths or brings hidden things to light.

Sexual barriers should be totally obliterated and these new attitudes translated into law.

These are some of the possibilities. The reader will perhaps be able to contribute even further prognostications from his own unique view of Scorpio-Taurus.

****

PLUTO FUNCTIONS	Taurus-Scorpio	Gemini-Sagittarius	Cancer-Capricorn	Leo-Aquarius	Virgo-Pisces
ANNIHILATION, MASS DESTRUCTION	Civil War	Spanish-American War	W.W.I.	W.W.II Korean War	Vietnam War
ACHIEVEMENT, UNIVERSAL WELFARE	Labor Unions	Public School System	Social Security	United Nations	Medicare
MIDDLE TURNING POINT, CRISIS	Johnson's Impeachment	Spanish American War	Wall St. Crash	Atomic Bomb	Vietnam War
POLARITY EMPHASIS	Reconstruction Era	Wildlife Conservation	Depression	Scientific Advances	New Frontiers in Mysticism
GROUP ACTIVITY, MOVEMENT	Agriculture, Industry	Migration, Immigration	Woman's Suffrage	Dictators, Monarchies Abolished	Ecology
NON-CORRELATING EVENT	Transcontinental Railroad				Moon Landing, Women's Liberation Movement
TRANSFORMATION	Banking System	Communication, Transportation	Public Welfare Agencies	Science	
REGIMENTATION, STANDARDIZATION	Uniform Currency	Standard Time Zones	Prohibition	Totalitarianism	Computerization
EXTREMISM, BREAKING BARRIERS	Millionaires	Theory of Relativity		Atomic Energy	Drug Addiction

SOME MAJOR EVENTS SUMMARIZED

HISTORICAL SUMMARY

It is obviously difficult to categorize complex situations and events which cover the range of human affairs. Nevertheless, there is a broad pattern which can be detected even though we readily acknowledge that some of these assignments could be equally applicable to other departments. The establishment of the public school system, for instance, is at once an achievement in the interests of universal welfare, a group movement, a transformation of the existing order, and certainly a form of regimentation. This, however, only enhances the astrological view and is beside the point.

The issue is whether the qualities we attribute to Pluto did or did not manifest by sign transit, and a glance at the chart will support the premise in most instances, thus leading us to some fairly reliable conclusions:

1. Pluto was operative long before we were aware of its existence. Therefore, we are undoubtedly being influenced by forces of which we are as yet unaware.

2. Our wars and annihilative forces tally very well with Pluto's transit, and we should be aware of this relationship so as to avoid further conflict.

3. Pluto demonstrates its power regularly in the area of universal welfare, which belies the outworn theory that its function is primarily negative.

4. The midpoint of the Pluto era is likely to be unusually significant, thus supporting the belief in the importance of midpoints.

5. The polarity sign manifests very distinctly and we would perhaps do well to examine this potential more often in terms of natal interpretation.

6. The group action which seems to accompany Pluto's transit is simply a result of many people being similarly motivated at one time. Groups are still made up of individuals, and we can ensure a positive direction for this powerful force by being aware, individually, of the alternatives involved.

7. The extremist quality of Pluto's action can provoke exceptional performance in "one-out-of-many". The status of millionaire was achieved by few, and the theory of relativity conceived by one man. Let's abandon the notion once and for all that Pluto operates only on the

impersonal group level.

8. What is promoted, invoked or enacted under Pluto's tutelage has an endurance which almost defies extinction. This is particularly relevant in terms of groups founded or laws enacted; what is established becomes "a way of life."

****

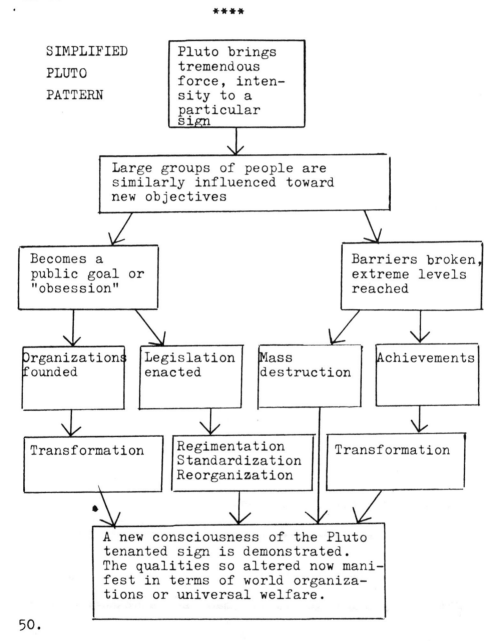

SIMPLIFIED
PLUTO
PATTERN

Pluto brings tremendous force, intensity to a particular sign

Large groups of people are similarly influenced toward new objectives

Becomes a public goal or "obsession"

Barriers broken, extreme levels reached

Organizations founded

Legislation enacted

Mass destruction

Achievements

Transformation

Regimentation Standardization Reorganization

Transformation

A new consciousness of the Pluto tenanted sign is demonstrated. The qualities so altered now manifest in terms of world organizations or universal welfare.

THE PLUTO PRINCIPLE

The foregoing historical overview obviously supports the belief that the Plutonian expression is broad in scope, motivates large groups to co-ordinated action, and is enduring and dynamic by nature. This is, unfortunately, where many Astrologers stop, and at this point we are far short of the Pluto totality.

The fact that a man is called into military service along with thousands of others does not diminish for <u>him</u> the uniqueness of the experience. And you'd have a hard time convincing most people that because they pay income tax along with millions of others it has no significance for them personally! For some peculiar reason, this is the kind of thinking which has prevailed with Pluto, and it is altogether limited.

An apple is an apple, whether we are making apple pie, applesauce, or picking it fresh from the tree. Likewise, Pluto is Pluto, whether we are scanning national events, generation qualities, or the individual. We have been observing Pluto on a quantitative basis, but must now change our focus somewhat in order to bring its qualitative aspects into better view for natal interpretation.

When we learn a new language we can study by memorizing a certain few words, or learn the ABC's and thereby read <u>any</u> word. The following is a suggested assignment of the Pluto "alphabet"; when principles are understood the specifics will follow.

Pluto Intensifies

Wherever Pluto resides we find a distinctly concentrated situation, It is a "much-more-so" configuration. If the indications are those of strength, then the individual becomes, through Pluto, <u>much</u> stronger....if of weakness, <u>much</u> weaker, etc.

Pluto is, of itself, neutral; it simply lends to any person or situation the unique ability to carry on further than the usual limits. This is sometimes easier to comprehend in terms of extremes, since Pluto is certainly the arch extremist.

Pluto operates most always outside the limits of normality. Think of the normal bounds of any situation, then move beyond....<u>now</u> you have reached the Pluto realm.

This is the underlying principle for Pluto's relationship to anything "fantastic" or "miraculous". His modus-
operandi is simply outside the norm. He is "hyper", "hy-
po", "sub" or "super", etc.

Saturn square Venus from the Fourth House, for instance,
can indicate a poor or humble childhood home, but let
Pluto get involved in the situation and you can find extreme poverty, or a penitentiary atmosphere. Or take Jupiter trine Mars from the Ascendent; a configuration of an
extremely healthy, vibrant body. When Pluto chimes in you
may well find a first-rate athlete, possibly capable of
olympic status.

The first step, then, is to move out of the ordinary and
into the realm of the <u>extraordinary</u>.

Pluto Standardizes

We must remember that many people are born with Pluto in
the same sign and position, so that what is individually
extraordinary is also standard equipment for a large seg-
ment of the population at any given time.

If we could somehow bring back one person with Pluto in
Aries, we would likely detect a singular ability for
leadership or pioneering. At the time such a condition
prevailed, however, this tendency must have been considered commonplace.

When 50% of a class receives an "A" on an exam, it seems
somehow to diminish the distinction, but we should still
recognize this mark of excellence. Pluto is comparable
to just this type of achievement: a great many have arrived at the same time who have reached a particular level
of accomplishment in some similar respect. Astrologers
who claim that only certain highly evolved individuals
can respond to the better vibrations of Pluto have perhaps overlooked extraordinary ability simply because it
has become temporarily common. The qualities existent in
any generation are only exhibited, after all, since they
manifest in each and every member of that group!

Pluto Breaks Barriers

This is another facet of the extremist nature of Pluto's
activity, but an underlying concept which is important
in understanding the Pluto principle.

The physical prowess, clairvoyance, lawlessness, etc. which are possible through Plutonian aspects are conferred for just this reason: Pluto crosses all the usual boundaries and breaks through normal barriers.

A strong configuration involving Pluto is a "record-breaking"situation. New levels beyond the norm can be reached; laws broken, quite literally, or the "law of averages" upset.

Recall for a moment how Pluto influenced our collective consciousness time and again, inciting us to cross the geographical boundaries of our land and march off to war; or how, through the impact of an Aquarian influence we bypassed the barriers which separate nations and established the United Nations. Walls between countries or individuals are virtually wiped out, and the result is an unsurpassed ability to unite.

Many Pluto aspects thus imply a universal quality. A Pluto-Aquarius or eleventh house emphasis enables one to mix and mingle with all types. A Pluto-Cancer or fourth house emphasis depicts one who makes the world his home and is at home in the world.

Consider the usual barriers or limitations in a given situation, then visualize these barriers as being pushed aside; this is Pluto in action.

Pluto Separates

Any time exceptional departure from the norm is indicated, as it is with Pluto, one is automatically cut off from the main body of individuals who are operating within the ordinary context. In this sense, then, there is most always some form of separation evident.

Take, for example, a person who "mothers" every child on the block: the energies are so all-encompassing and so diffuse that it is impossible to dote on just one or two children. The nurturing bent has departed from the limited norm, and has taken on a universal quality. Or take the individual to whom everyone and anyone is "friend": there is little time or energy left to develop one or two close "buddy" relationships. This is the way Pluto operates; separation from the usual or ordinary circle of activity, but functioning in a larger or universal sphere.

Pluto Reaches Extremes

Does this mean that each and every person is an extremist in one respect or another? Not necessarily. However,
even given a very balanced personality not given to exceptional display generally, the individual should <u>still</u>
be able to extend beyond the normal limits in some area.

Some individuals are, of course, by their very nature
more radical than others, so that such a quality will be
immediately more obvious. However, if we look beyond an
undistinguished front some unusual degree of prowess
can often be detected.

This Pluto-related ability does not indicate a point for
necessary development usually, but simply an innate
quality which must, as always, be properly channeled.
The power is already existent, and needs only to be
drawn upon. This is a money-in-the-bank situation, or
already well-developed muscle; thus the tendency to
reach unusual heights or depths.

Notice, for instance, those who have Pluto on the ascendent or the MC: you can easily detect a power-wielding
force at work, no matter how deceptively gentle the
total personality may be.

Pluto Represents One-Out-Of-Many

The very uniqueness and universality of the Pluto expression brings into being a "one-out-of-many" situation.
Those who break through the normal barriers and limitations, thus separating themselves from the group by
reaching some extreme level are bound to be the minority,
or the "one".

What is the distinction, then, between Uranian individuality and Plutonian individuality? There are similari-
ties in that both stand aside from the group, but the
Plutonian retains his membership, while the Uranian
does not. Uranus in the 7th, for example, most often
implies a split in the partnership, and many times no
marriage at all in the traditional sense. The Pluto type,
however, can be partners with all; have many marital
arrangements, or one in which the partners stay together
yet lead two separate lives. The Uranian is marching to
a different drummer and may even insist on starting his
own parade. The Plutonian is simply marching better
than, different than, or farther behind anyone else. He
is still a member of the "band."

54.

Pluto Reaches Goals

Because of the innate strength and power indicated by the Pluto position and aspects, an individual who seeks attainment in that direction will find the going relatively easy.

This is an available source of power which should not be taken lightly, since any efforts put forth here will be rewarded quickly and forcefully.

A Pluto-Neptune aspect, for instance, will enable one to inspire others to an awesome degree, and hopefully the sense of responsibility is equal to the task. Otherwise, there can be trouble on a grand scale such as in the case of Adolf Hitler who had Pluto conjunct Neptune in the Eighth House, and was able to inspire a whole nation to "do or die" for the wrong reasons.

****

THE PLUTO GENERATIONS

> 'If there is any responsibility in the cycle of
> life it must be that one generation owes to the
> next that strength by which it can come to face
> ultimate concerns in its own way"
>
> Erik H. Erikson[1].

Since Pluto usually has a 20-30 year sign orbit, it is
sometimes called the "generation" planet. At the time of
transit, as previously noted, tremendously important
events occur which are timely to the sign which Pluto
tenants. At the same time, huge numbers of people are
born who also share this common Pluto bond. These are
the souls who truly understand what all the "fuss" was
about and are in fact the personification of that era
and those principles.

It is very much like planting a garden. The "seeds" are
planted during the Pluto era, but the blossoming or mat-
uration period arrives much later. The growth, manifestation, or blossoming is evident, even though by now
much attention is directed to "planting a new crop" or
dealing with new issues.

The upheaval, stress, and sometimes destructive tendencies which emerge during the Pluto transit are easy bag-
gage for the souls born at the same time. Having already
incorporated such qualities into the ego, and developed
them to a high degree, they are drawn to a time when such
is being emphasized. If solutions are needed, look to
these people, for what we currently ponder, they already
know.

Let's examine the current assortment.

Pluto in Gemini Generation

Born: 1882-1914   Maturity: 1902-1934

This particular group arrived in a rather dark world,
literally speaking. It was the changeover period, from
the agrarian to the urban society, and as noted, a time
of tremendous emphasis on new methods of communication.
Perhaps no generation in history has witnessed such
vast change; a world transformed from the horse and
buggy days to the space age. They have taken it all with

a certain aplomb, adjusting readily to each new fantastic change as it came along; eager for progress and re-
ceptive to new thought.

Could any group have done as well? I think that the Pluto
in Taurus people would have had trouble in this era, and
no doubt stood smack in the way of any progress. These
Gemini souls arrived at the appropriate time to move us
all along a little. When the automobile, airplane, and
radio were introduced they accepted the premise and made
it a reality.

This is not to say that there were no traditionalists
among their ranks, or that there are no adventurous spirits elsewhere; rather that as a group, they _could_ have
rejected the new trends, in which case we might all still
be on horseback. Pluto in Gemini brought us the new communicative tools, and the Pluto in Gemini generation fos-
tered the communicative spirit.

You can almost anticipate a certain dilemma as Pluto
transits a particular sign, and a somewhat awkward attempt to find expression or solutions. Watch, as the
group born of this particular Pluto consciousness matures;
it is as if they were saying, "Let us show you how to do
it"....and count on them doing just that.

Pluto in Cancer Generation

Born: 1912-1939   Maturity: 1932-1954

As each group has its special talents, it also manifests
its own specific weaknesses. The adventurous, devil-may
care attitude of the Pluto in Gemini group was possibly
part of the problem which led us to a national financial
crisis, the depression of the 30's. The situation was met
head-on at the time; it had to be. But if you want to see
how real security is created, look to the generation of
that period, for they are inherently security-oriented.

Here we find those known as "The Silent Majority". While
the world whirls on its merry way, this group has quietly revolutionized the meaning of home and home life. They
have built "nests" by the millions, feathered their nests
("consumerism"), and paid the bills for the largest group
of youth and elderly in our history.

Though patriotism was at an all-time high during the
First World War, there are probably more who gave their
lives in name of country from this particular group

(World War II, Korea) than from any other. And all this without much fanfare, since love of country is absolutely ingrained in the Pluto in Cancer consciousness.

These are not necessarily the best parents ever to hit the world scene, but I do think that never before has any group tried so hard. While the Pluto in Gemini parents were certainly not the dominating types of the Taurus era, they were not child-oriented or caught up with the concerns of children; quite a contrast with the Cancer group. Babysitting emerged as a profession, and many women worked both in and out of the home, revolutionizing the concept of home itself. Yet never before was there so much interest or emphasis on the needs, both material and psychological, of children.

The Cancer expression is not typically authority-oriented, and ultimately there was a breakdown in family unity and discipline. However, where leadership is concerned, there is no one more capable than a Pluto in Leo type, and so the torch is passed.

Pluto in Leo Generation

Born: 1937-1958   Maturity: 1957-1978

This is a high-minded and masterful group, with a will to dominate others, and they started early with an easy target: their Pluto in Cancer parents.

Pluto entered Leo in 1937 so this group came to maturity in 1957, approximately the time all the trouble started on the campus and homefront. They have defied their parents, their government, and their society; at the same time demonstrating uncommon ability toward leadership. They have questioned traditional values, and in many instances persuaded the entire country that they were right to do so. Standards of marriage, family, patriotism, education, etc. which had been blindly accepted by others were questioned by these youth, and in some cases changed so that our social directions will never be quite the same again. That these types were able to accomplish so much, so early, should be a large clue of what is yet to come.

The Pluto in Leo generation has an inherent grasp of the meaning of power and authority. They instinctively know how to gain their ends, and used wisely, this ability could move mountains for the eventual betterment of everyone. We can expect some really spectacular leaders from these ranks, but we have to watch for the "Little

Caesars", for they too are waiting in the wings.

While we pass quickly on to new challenges, these people will continue to revolutionize the meaning of love; also an aspect of the Leo consciousness; and the meaning of authority.

With all this dramatic thrashing about in marches, promoting first one cause and then another, the practical necessities of our work-a-day world must still be dealt with, and for this we pass to the Pluto in Virgo generation.

Pluto in Virgo Generation

Born: 1956-1972   Maturity: 1976-1992

Pluto entered Virgo in 1956, so these types won't begin to emerge as a group force until about 1976. While we struggle valiantly with the problems of ecology, these folks will be able to step forth with some really tenable solutions and show us how the big clean-up is to be done.

Virgo is meticulous, mathematical, and generally scientific, so these people will also be able to better handle the computerization that is swamping the rest of us. They may also, through this scientific know-how, revolutionize the concept of work itself. Efficient methods may cut our work time further still, as the Virgo consciousness swings into action.

Health standards should be a major concern of this particular group, and if it doesn't come about before, we can expect socialized medicine to emerge as a reality when these people mature. As voting adults, the Pluto in Virgo generation will undoubtedly support any national plan for physical well-being.

Virgo is a tireless worker, along with being something of a perfectionist, so some really admirable craftsmanship will return to our industry, and humble work will be returned to a position of due respectability. The white-collar vs. blue-collar notion which has prevailed with a slightly snobbish Cancer-Capricorn group will be left in the dust by Pluto in Virgo.

While the Cancer Generation directed its energies toward keeping the wolf from the door, and the Leo Generation geared itself toward worldly and humanitarian issues, the Virgo Generation will demonstrate an uncommon ability to deal with the nitty-gritty problems of everyday life.

Pluto in Libra Generation

Born: 1971-1983   Maturity: 1991-2203

Libra is sometimes accused of being lazy, indecisive, or even wishy-washy, but to their credit, they are rarely war-like. The Pluto in Libra generation should have an inherent grasp of the cooperative spirit, so necessary for harmony in all relationships. The modus operandi is that which is harmonious, fair, and just; a consciousness directed toward "us" or "we" and therefore the probability of rather peaceful worldly conditions.

While the Virgo Generation may have become bogged down with practical, utilitarian necessities, the Libran group will be able to add some beautifying touches. Artistry and appreciation for beauty are strongly accented in Libra types, so we can expect some exceptional improvements in architecture, industrial design, etc.

The pioneering spirit will demonstrate itself through the Aries polarity, and these types would be willing candidates for space or planetary emigration.

This generation will exhibit a disinclination toward sordid or dull matters, as well as progressivism, artistry, and a well developed social conscience.

****

The Plutonian principle of standardization emerges through the Pluto sign or generation link. This is the group to which we belong as social beings; the kindred spirits with whom we share a particular orientation in terms of goals, principles, and directions.

60.

PLUTO THROUGH THE HOUSES

It is very difficult, if not downright impossible, to take placements and configurations out of context. Each chart has its own particular qualifying factors to consider, and here we are attempting to isolate that which in reality is never isolated.

A good doctor never examines part of the body without consideration of the whole; yet we do just that when we analyze bits and pieces of astrological lore. This burdens the beginner in particular, since he is inclined to mistake an "ideal" or "typical" interpretation for The Gospel.

Still, we must examine the pieces of the puzzle separately before we can properly put them together, and the following is offered on that basis only. These are examples, and may be greatly altered by individual contradictions.

Pluto In The First House - Example

Since the first house relates primarily to the body and superficial personality, this Pluto position (or conjunct ascendant from the 12th) makes the individual himself the personification of the Plutonian traits.

Pluto carries the Scorpionic tendency to secretiveness, and this person can appear rather quiet and aloof; but never confuse "quiet" with "docile"! This type is intense, determined, invincible, unyielding, dauntless, and frequently dictatorial. He is likely to be a tower of strength to those around him, and a good person to confide in, as he is the exact opposite of the chattering, gossipy type. He may or may not (usually not) have warmth of personality, is rarely weak unless many other contributing factors prevail, but is always "different".

He is a law unto himself and may, in fact, be "lawless". If he operates within the bounds of society he is still his own person through a self-imposed separation from tradition. He can go through the motions, but remain absolutely unchanging.

Relative to the physical body, an unusual sturdiness may be evident, and in many cases extreme physical

prowess. If extremely afflicted, it may denote some
physical deformity since this too is an extreme.
If other factors support it, the physical condition may
also lend itself to advanced psychism. The total picture
has to be evaluated, but do look for some manifestation
of hyper-development, physically speaking.

Pluto In The Second House - Example

Here again we look for the extremes; some uncommon ability, attitude, or interest in the Taurean or second
house areas. The most common interpretation is money,
per se, and you might find anything from a thief, to
someone with an unusual talent for stretching the dollar,
all the way to the millionaire. In some way he stands
out or excels in this department. Unless other factors
contradict, this type is also strongly materialistic.
He has an instinct for self-preservation and an ability
in general to amass, collect, or to build up his assets.
This is excellent, of course, for bankers, businessmen,
and others dealing directly with money. The money sense
is innate, the judgment thereof keen.

Since Pluto pushes past barriers however, this individual may push aside monetary principles altogether, ex-
changing materialism for "naturalism" or a life devoid
of the usual natural resource, money . In some way,the
step by step road to riches is bypassed; he jumps right
to the top or disdains it altogether. The Taurean second
house represents the supremely natural, and this is the
important key here. This person is comfortable in the
earthly pursuits: he is in the world and of the world,
and only rarely a "soul-searching"type.

Every configuration has its negative connotations, and
Pluto in the second "takes what he wants", thus the
thief, extortionist, clever manipulator, etc. As always,
expect some exceptional development.

Pluto In The Third House - Example

When Pluto highlights the Gemini third house qualities,
we find extreme ability in areas of communicativeness,
wit, dexterity, and frequently (for good or ill) an uncommon mentality.

This type is able to mix, on a superficial basis, with
all and sundry (all air signs, in fact, have this ability to some degree). He is the communicator, par
62.

excellence with pronounced oratorical ability. As such,
he makes a marvelous salesman, teacher, radio or TV
announcer, etc. One case comes to mind here; a man with
Pluto in the 3rd with six aspects; he was considered a
"super-salesman"

The mind is clear, sharp, and clever, and if other factors support, may denote, through the "barrier-breaking"
Pluto action, some particularly uncommon bent: ESP, mind-
reading ability, or even some derangement. He can even
be uncommonly good at duping or decieving others; a
swindler, trickster or "con" man.

Versatility and dexterity should be quite evident on
either a physical or mental basis. This is a "Jack-of-
all-trades-master-of-several" configuration; the "handy-
man" or well rounded individual.

Since the 3rd house also represents one's environment,
the Plutonian emphasis can signify one who functions
extraordinarily well on a day to day basis, with a faculty for making daily routine an exciting, significant
experience. What can be dreary, ordinary events to many
become, to this type, an adventure.

Pluto In The Fourth House - Example

When Pluto dominates from the Cancer fourth house domain,
all the traditional concepts of hearth and home are a-
bandoned. "Home" is anywhere and everywhere. If there are
strong counter-indications of a conventional sort, the
person will _still_ exhibit an uncommon ability to be comfortable almost anyplace. There are no walls, remember,
within the Plutonian framework.

All angular positions are unusually important, and Pluto
on the 4th angle denotes a situation of extreme soul-
searching. The inner life is a rich one, with tremendous
intensity directed toward inner growth. The experiences
of this incarnation offer an unusual chance for spirit-
ual development due to this emphasis on sensitivity, and
the person in fact often becomes hyper-sensitive. If
they seem "touchy" or easily offended, it is simply this
ultra-quick reaction pattern in action.

The nurturing qualities are somewhat out of bounds, and
the individual may well act as parent to all while immediate family is neglected. The scope of the small family circle is simply too small to contain the Plutonian
action.

Relative to the nurturing function, it is interesting to note that Louis Pasteur had Pluto in the 4th and his research with milk brought safer standards for the entire world; Pluto once again demonstrates its universality.

Early separation from the parents and/or family is also a distinct possibility when Pluto takes this position. It may be actual or symbolic, but some form of estrangement or distant relationship is indicated.

This is an excellent position for people who deal in real-estate, home-building, home-making, etc., since it denotes extraordinary ability along these and all Cancer related lines.

Pluto In The Fifth House - Example

When extremist Pluto resides in the fifth house, one is inclined to have few children, or none at all.(Naturally, both prospective parents must be considered in this regard.) The individual may be separated or estranged from his (her) own children, yet can relate well to children generally. Thus, this type makes a good teacher, counselor or foster-parent. They are capable of becoming absolutely obsessed with youngsters and/or their problems.

The usual romantic episodes prior to marriage are few or none. The individual marries the "first love" or has one truly exceptional love affair. This type is also capable of many strong and intense relationships, so that afflicted Pluto in the fifth may have one affair after the other, after marriage.

This person has the unique capacity of being able to enjoy anything and everything to some extent; he has an appreciation of life in all its aspects from the ridiculous to the sublime and so will often have many and varied hobbies, at least one of which is sure to be unusual.

Some unusual degree of energy will undoubtedly be directed toward creative expression, since this type is inherently creative. It may be channeled through the fine arts, dancing, acting, etc., but some outlet is generally found since the need for self expression is great.

Pluto In The Sixth House - Example

If Pluto resides in the Virgo sixth house, you can expect to find some unusual turn of events connected with
health or daily routine.

Since Pluto's residence indicates a bodily concern which
is out of bounds, this can be a prime indicator for
hypochondria. Of course, the individual may in fact be
plagued with some condition which could be termed "extraordinary"; judgment is called for here in determining
the specific case. Either way, the individual is em-
broiled in issues of health and well-being. Since this
tendency is innate, such a configuration is naturally
of great assistance toward accomplishment in any of the
medical professions, with the degree of ability being
of a generally superior caliber.

Negative Pluto can be sickly or chronically out of work,
but this position,well aspected,points to a good solid
worker; one with persistence and stamina and a high
degree of development; mastery, if you like, in the area
of good old daily routine.

A fine example of Pluto extremism at work in this department is depicted by Mahatma Ghandi who had Pluto in
the sixth house. In utilizing this over-developed "muscle", the Pluto principle was put to work in a positive
direction: through fasting Ghandi helped to free India.

Pluto In The Seventh House - Example

The 7th house Pluto position indicates a consciousness
which has been overdeveloped in terms of "the other
person", or partners in general. This is the essence of
the Libran team spirit, and denotes one who will bend
over backwards to cooperate. Without balance for this
tendency, an individual could become overly dependent
and short on his own ego strength; a "tell me what to
do next", or "what do you think" approach to life.

This Pluto position also indicates a keen insight into
just what makes other tick. They know how to motivate,
how to relate, and have the social know-how to do it
diplomatically. This is excellent too, for business partnerships, social work, and mediators of all kinds. A
certain refinement can also be noted: I have yet to
meet anyone with this configuration who was in any way
crude, and the esthetic-artistic sense is well developed.

65.

Pluto in the 7th simply can't be expressed in the usual sense regarding marriage. These types relate readily but not necessarily deeply to others, and so may take one partner after another _or_ establish a relationship wherein each is free to go his own way. The marriage partner is usually dark in appearance, strong, dominating, and wilful

Pluto In The Eighth House - Example

Pluto is quite comfortable here in its own domain, and a well-aspected Pluto in this house may well indicate super-natural occult ability. The 8th represents "other states of consciousness", and a power source at this point is a great impetus toward the occult arts. Clair-voyance, black or white magic, mediumship, states of trance, etc. are all possible through Pluto-power. Even in those who are reasonably materialistic or earthy, the degree of sensitivity will be highly developed.

Pluto in the 8th also indicates a highly developed will-power. These types are quite indomitable. They make fierce adversaries and simply refuse to give up, regard-less of the odds. This quality lends itself well, of course, to any martial career, the law, or any profession which requires a superior degree of persistence and ag-gression. Scorpio is one of our scientific signs, so the 8th emphasis can also be pertinent to any of those pursuits, especially medicine and some of the new branches of ecology. Detecting and therefore detectives are also blessed by this configuration, since Pluto's ability is shrewd, clever, sometimes even conniving, but always perceptive in the investigative pursuits.

Pluto In The Ninth House - Example

When extremist Pluto takes up residence in the Sagittar-ian 9th house you may find a first-rate lawbreaker, a religious fanatic,teacher,philosopher, mystic, or even some sort of genius.

One person comes to mind who encompasses most of these possibilities: this man taught religion and philosophy and was considered brilliant. He was completely lacking in moral fiber, however, so ran afoul of the law and did spend some time in jail. Pluto out of bounds again. If the ethical nature had also been well developed, the individual could have arrived at a level of some distinc-tion. Pluto by itself, powerful as it is, simply isn't

66.

enough to carry the life load. It _is_ extreme power, but is not _all_ _powerful_.

Religious devotion of the very highest order is also possible with this position _or_ you may find the religious fanatic.

If the Ascendant is supportive, and the chart in general leads to a physical emphasis, this position can also indicate achievement in sports, although Sagittarius generally leans toward outdoor activities, and only rarely "games."

Pluto In The Tenth House

Pluto in the Capricorn 10th house is in its most powerful position, especially if conjunct the Midheaven. Because of its tendency to act extreme, it is most important that it be used wisely in this instance. Any planet on the Midheaven, in fact, sort of "rules the life", so watch that planet.

This is the hallmark of those who are driven toward accomplishment. They just keep climbing and climbing, ever thirsting for further success. They drive themselves and everyone around them, so if you're not especially ambitious you'd better laze around somewhere else.

This is probably one of the very best configurations for success, prestige, accomplishment, and recognition. It denotes many lives of effort toward those ends, and the individual has an instinctive ability to climb to the top of the heap. ('How To Succeed In Business While Really Trying'). This is excellent for political know-how, business in general, and management in particular.

A better worker would be hard to find (maybe Pluto in the 6th), and the plain old "horse-sense" conferred here is also unsurpassed.

Out of bounds Pluto in the 10th is the political schemer, determined to serve his own ends through any means, or the crafty businessman who climbs on anyone and everyone to get to the top.

A superior leader, a wise merchant, the community-oriented worker, the dictator, and political schemer are all possibilities.

Pluto In The Eleventh House - Example

The individual who has Pluto in the Aquarian 11th house
will usually demonstrate an uncanny knack for making
friends. The separative activity of Pluto tends to keep
the relationships somewhat impersonal, however, and close
bonds are abandoned in favor of the universal approach.
Racial, social, age, and sex barriers are all bypassed,
and relationships evolve which are genuinely friendly
but detached.

This is an excellent configuration for those who work
in the social sciences: psychologists, sociologists, and
those who deal with people generally, since they are
endowed, through the Pluto position, with scientific
detachment and a genuine concern for others.

Negative or out-of-bounds Pluto can become a little too
detached or independent here, insisting on having and
going his own way; and if Pluto is badly afflicted, you
may suspect a really off-beat or anti-social individual.
Afflictions can also indicate sudden and sharp endings
to friendships and/or a truly strange social life.

The inventive faculty is likely to be strong, and if
Mercury assists, some unusually innovative ideas can surface. All scientific endeavors, in fact, are well supported by this placement.

Progressive reformers, scientists (especially social),
teachers, and extreme non-conformers are appropriate for
this configuration.

Pluto In The Twelfth House - Example

Pluto in the Piscean 12th House gives super-normal ability
for any of the Piscean-related occupations: research, in-
stitutional work of all kinds, psychic endeavors, etc.
The tendency here is to withdraw, but for powerful pur-
poses. There is a great deal of strength behind the
scenes and if nothing else, the individual will have un-
suspected resources at his disposal; "hidden" power.

This position also frequently indicates super-normal kar-
ma, and the person may partake of the most fantastic and
unusual experiences imaginable. He chose, evidently, to
dispose of quite a few obligations at one fell swoop.

Afflicted Pluto can indicate loneliness to an almost
overwhelming degree. The 12th represents isolation, and
Pluto makes the situation extreme.

68.

Extreme sensitivity, depth, and a life filled with inner experience are all indicated. As with Pluto in the fourth, hypersensitivity is possible, and thus is a supportive condition for anyone developing along psychic lines.

There is some departure from the norm in one's ability to understand; afflictions to Mercury, for instance, can indicate something as basic as a language barrier; or a well aspected Pluto can point to one who is truly evolved in terms of compassion and depth of understanding.

Poets, artists, psychics, counselors, researchers, or those who live lives withdrawn from society are all possible through Pluto in the 12th.

****

The Plutonian principle of extremism is demonstrated through and indicated by house placement.

Pluto intensifies...

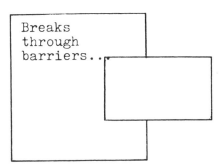

Breaks through barriers...

Separates from norm; reaches extremes

PLUTO IN ASPECT

Some Astrologers view squares and oppositions as negative, trines and sextiles as positive, and conjunctions
as neutral. Others maintain that the positive or negative quality depends on the planets involved, regardless
of the configuration. Since the former view is more
widely accepted, the following examples have been noted
accordingly.

Pluto Conjunctions - Examples

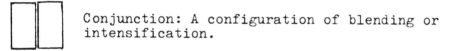

Conjunction: A configuration of blending or
intensification.

SUN - The individual is stable in the extreme. The inner
nature is such that outer circumstances, events, or
opinions do not in the least affect the core person. An
inner calm prevails in circumstances which would unsettle
less stalwart souls and unless there are severe afflictions to the nervous system, these types can be noted
first and foremost as being serene. Inwardly they are
utterly assured, and sure of their own directions, goals,
and convictions. There may or may not be a contradictory
facade (Moon or Ascendant of contradicting type), but do
not be misled: this type goes his own way according to
his own lights. He does not change easily, and under
severe pressure will simply become immobilized. While
Pluto on the Ascendant gives the impression of being a
tower of strength, this type is a tower of strength. You
can chip away granite, friends, but this aspect is akin
to the diamond.

MOON - Heredity patterns or environment have conditioned
this individual toward stability, control, composure,
and a personality which is perceptive, dominating, and
forceful. He is deeply emotional and may even have an
unusual amount of inner turmoil; nevertheless, he is
well controlled and may even appear to be "cool". When
he cares, he cares very much; when he doesn't, not even
an act of Congress can change his feelings. He will give
completely of himself of his own volition (never yours),
and his total dedication and zeal is unmatched. This is
a private person, not at all easy to know or understand,
70.

but he will not pass through life unnoticed.

MERCURY - The Plutonian mentality is secretive, non-confiding, rational and argumentative. These types are mental aggressors and with supportive aspects to this configuration are also capable of unequalled logic and thus make excellent debaters. These individuals respond well to reason, but are rarely influenced by emotion (present the facts and forget the tears.) They are rather certain of the validity of their own opinions, and rarely change once the mind is made up. Very little escapes the Plutonian "eye" and these types also exhibit an excellent faculty for detail. Above all, look for some exceptional quality connected with the mind.

VENUS - This type of love nature is loyal unto death. What they love, they love, and who they love, they adore. This is a quiet, enduring, sustaining sort of aspect, completely devoid of gushy sentimentality. Expect birthday presents and warm responses from someone else; Pluto-Venus types can't concern themselves with trivia. If you ever need a friend , however, and you happen to be half way around the world, this is the one to call; he'll be there! By the same token, any ill will engendered can finish a relationship forever. When Pluto is involved we are dealing with black and white situations; there are no "greys" and through a rebuffed Venus, no forgiving.

MARS - These types have strength, drive, and a competitive spirit unmatched by any other aspect (except other Pluto-Mars). The physical drives are strong, and often imply an unusual interest in sports, games, and all activities of an earthy nature. This denotes an individual who never gives up, no matter the odds. These types have a great deal of "push" power and know exactly how to use it. The house position, of course, must be noted to determine exactly where the native will put this to use. There is also some undue tendency to anger, and hopefully the individual will find a proper outlet for his strong feelings. This is an aspect of tremendous energy and pure power.

JUPITER - Instantaneous, perceptive, and shrewd judgment is depicted through this aspect; excellent, of course, for those whose work requires quick and accurate decisions. This represents sage, sound advice with no frills attached. The Pluto-Jupiter type is also capable of an unbounded giving of self in some area (see house position), though as always with Pluto involvement, it is

71.

self-determined. There should also be numerous oppor-
tunities foracquiring those things which lead to "the
good life". Unless extreme afflictions prevail from
other sources, the individual is always provided for,
or is able to provide for self.

SATURN - Pluto-Saturn types are capable of carrying
great responsibilities and are therefore usually found
loaded with same. These people are some of our most
reliable and best workers, going from one task to the
next with seemingly little effort. Other factors being
equal, they are also usually very big on integrity,
and can be absolutely uncompromising on questions of
morals or ethics. Unless severely afflicted, the Pluto-
Saturn person will be trustworthy in the extreme, and
will insist on doing the "right" thing no matter how
much (or who, for that matter) it hurts.

URANUS - This aspect represents a type who is very wilful and determined, and can demonstrate an almost un-
believable amount of persistence. If you will visualize
a cement block entrenched in cement, you've got the
picture. A certain cleverness also prevails, and when
thwarted, this individual will simply devise new and
better ways of achieving the same purpose. Once the mis-
sion is accomplished, he is off to some new quest, and
once again persists until he is successful. Scientific
or mechanical talents may also accompany this aspect,
as well as an individualistic orientation to life.

NEPTUNE - Idealistic fervor typifies this aspect, along
with the ability to inspire and lead others. Excellent,
of course, for salesmen, religious leaders, actors, etc.
This type can also elevate some quality to the level of
an art (note house position), and may in fact exhibit
a talent for one of the arts _per se_. The perceptions are
keen, the intuitive faculty well developed, and the individual is, generally, super-sensitive. While this type
may at times seem perplexing to others who do not quite
fathom his peculiar or mysterious ways, he understands
them only too well. When afflicted, there is no one
more deceptive or tricky than Pluto-Neptune; when well
aspected he is in tune with that which is sublime and
divine.

Pluto Trines and Sextiles - Examples

Trine: A configuration of a "given" quality.

Sextile: A configuration of an opportunity to develop the quality so indicated.

SUN - When Pluto and the Sun unite harmoniously, we can be assured that the individual so energized has learned somewhere, somehow the value and use of the power principle. He has an inner assurance born of circumstances which have taught him to stand alone, relying on no one. Because of this highly developed inner strength, he usually surfaces to the top in any group, the acknowledged leader, capable of taking command of most any situation (and he usually does). There is generally a strong bond with the father who was or is always supportive and probably quite dominating. These are the types whose convictions and actions need no sanction from others and whose inner directives are the compelling force. They are quietly forceful and subtly aggressive individuals who know what they want and go after it in an indirect but compelling manner.

MOON - These types are given an extreme amount of attention, concern, and in fact "doted" upon. Help from and through the mother can be counted upon, as Pluto trine the Moon mother-symbol signifies one who will always sustain, maintain, and to some extent dominate the life; the bonds are tight and are rarely broken. Extreme emotionalism along with the ability to control emotional reactions in others also accompanies this position; thus the individual may excel in acting, public speaking, or any field where meeting the public is paramount. Attention comes easily to the Pluto/Moon person, so it behooves him to be quite sure that any actions or activities can withstand the glare of publicity.

MERCURY - Many changes, travels and varieties of experience mark this aspect, and the individual so endowed is thus broadened accordingly. He is the epitome of the adaptable type, able to fit into any group or social situation with equal ease. Though he may or may not be the talkative type, he can, if necessary, out-talk anyone on almost any subject. This mentality is also the mentally investigative type, and can zero in on answers with such lightning speed that they are unaware

of the thinking process; they just "know", and they
know very quickly. These individuals are excellent at
repartee, and often exhibit a truly sharp and keen wit.

VENUS - Pluto/Venus types come from friendly, sociable
environments and are thus able to get along with anyone
and everyone. They make no social distinctions and so
cross all barriers with ease. Adult or child, rich or
poor, black or white, male or female; Pluto can charm
them all. It is not, however, of the overt or ingratia-
ting variety and the individual so aspected is much more
likely to seem rather quiet and aloof, yet still uni-
versally appealing. (FDR had this aspect which undoubtedly accounted in large measure for his mass appeal.)
Marriage and love relationships are also stabilized
through the Pluto involvement and though rarely close,
are of such enduring and sustaining quality that they
weather the storms which can cause others to fall by
the wayside. Pluto/Venus is love for everyone, and love
for a lifetime.

MARS - People with this aspect have somewhere, at some
time "earned their stripes" and so therefore rarely
show or display extreme anger or aggression. They understand fighting and violence with an inner awareness
that negates the need for a show of strength. "Mars the
brute" has reached, through Pluto, a new consciousness.
When and if the need arises for fighting or aggression,
it will be clever and covert, as this type can out-
maneuver almost anyone if he so chooses. An energy level
that almost defies extinction also is indicated, and
long after others have ended a weary day, Pluto/Mars
will be burning the midnight oil. A short sleep suffices,
and he is up at the crack of dawn.Endurance is the key
here, and if other factors support, a prime indicator
for longevity.

JUPITER - These are gregarious types whose good natured
optimism and even tempered disposition far surpass normal expectations. The sense of humor is also unequalled;
no matter how tense or how difficult the situation,
Pluto/Jupiter can be counted upon for some philosophical
humor which lightens the load for those who sometimes
fail to see the funny side of life. Good judgment is also to the fore, and these types can size up a person or
situation almost instantaneously and with uncanny accuracy.
Jupiterian abundance may be transcended through the
Pluto contact, and this individual either derives his
resources from a seemingly endless "horn of plenty"
through others, or is able to produce above and beyond
normal expectations.This talent for accummulation some-

times reaches such extremes in fact, that it becomes difficult to find a pathway through all the belongings.

SATURN - When Pluto joins Saturn in a harmonious configuration, we find a worker whose sense of duty and quest for perfection is unequalled. He may or may not be a haphazard type generally, but on the job he is dead serious; he faces his work with a sense of responsibility and great attention to detail. If and when any who are subservient to Pluto/Saturn receive praise, they can be assured that it is a rare accomplishment. This type is also without peer in his ability to discipline self. He can do without, go without, make do, and restrain himself when and if it becomes necessary. All the Saturnian qualities reach extremes through Pluto; above all, he is patient. Goals are set, meticulous plans are conceived, and the path is followed; it may take weeks, it may take years, but the end <u>will</u> be reached.

URANUS - The Plutonian/Uranian is likely to be a person in whom a new plateau has been reached in terms of individuality. He has gone beyond the need to be "different"; he <u>is</u> different. At some point in his personal evolution this type has been allowed to develop freely in his own way, and he no longer follows the crowd. He is likely to be out in front with the latest unconventional idea, and consequently attracts a great deal of attention from those who are less adventurous. Creativity and inventiveness loom large in these individuals. They can bring new and progressive ideas to stale situations; original solutions to old problems. These are the non-conformist reformers of our society; the people through whom Uranian eccentricity is channeled into new and constructive social change. This is an aspect of a restless, questing, impulsive <u>individual</u>.

NEPTUNE - People represented by this aspect are able to function unusually well in all areas which deal with intangibles; religion, psychism, the arts, etc. They have reached a point of exceptional development in Neptunian directions and are endowed with inherent artistry in some field as well as an ability to sense, feel, and deal with the abstract. These are the dreamers who are able, through Pluto, to translate dreams into realities. Fanciful or imaginative ideas are placed on canvas, religion becomes not only an inspiration but a pursuit, or the creative touch manifests as a new hair-do in the Pluto/Neptune beautician's hands. Pluto brings a rational, goal-oriented reality to Neptunian inspiration, and thus we find the dream made tangible. This is the visionary who also acts.

75.

Pluto Squares/Oppositions - Examples

Square: A configuration of that which is being earned or acquired

Opposition: A configuration of tension, awareness, or challenge

SUN - This individual is learning to properly express the power principle. He must develop a strength which comes from within and is not dependent on a "show" of authority: leading the pack, bossing the crew or whatever. He will be drawn to positions where executive ability can be demonstrated and should learn thereby that unrelenting, unswerving or dominating tactics do not work well. These types can learn from those whose evolutionary development is represented by the trine: authority is not dependent on dominating others, but domination of self; self reliance is possible only when the self is strengthened by independent action, and above all, true power comes always from within.

MOON - These types are learning important lessons in human relationships, and will generally be found in jobs dealing with the public. Pluto/Moon people are capable of pulling strings, clever maneuvering, and being too "pushy" in attaining their goals which tend to be self-serving. They must learn instead to seek accomplishments through friendly persuasion, sincerity, and concern for the welfare of others; also quite possible through Pluto-power. The challenge here is to discover how to elicit a desirable emotional response from others, instead of the emotional turmoil which usually manifests from those who feel"smothered" rather than "mothered", or taken advantage of rather than taken care of.

MERCURY - People represented by this configuration are learning to channel and direct all things denoted by Mercury: the mind, thinking process, movement, etc. There is a tendency to be mentally dominating and dogmatic or stubborn instead of firm; to be calculating, argumentative and shrewd instead of elevating the reasoning power; to flit from one project to the next instead of learning to adapt. These types are often found in the Mercury-related occupations or in situations which will provide the necessary discipline for proper development. They must learn to consider progressive ideas, to see another's point of view, and to adjust to new conditions and situations.

76.

VENUS - The Pluto/Venus type is found in situations
which will allow the development and refinement of the
social faculties. He is prone to "charm the birds out
of the trees" when it suits the purpose, yet the love
nature can be truly superficial and shallow. He must
learn to be nice because it makes life more pleasant,
and not just to achieve his goal. He must learn to
express love or friendliness when it is sincere, and
not simply to attain popularity. He must learn to curb
a tendency to be vulgar, since it has no place in the
Venusian realms of social interaction. These people
can learn from those who have the trine aspect; Pluto/
Venus must abandon the quest for universal popularity
and learn to express universal love.

MARS - These types have a tendency to use force and
guerrilla tactics to gain their ends. The need here is
to channel the aggressive expressions into constructive
rather than destructive activities; to build rather
than destroy. Pluto trine Mars has learned the futility
of brute strength, and it is this which must be acquired
by the "square" types. Some obsession is possible here
also,in terms of body-building in order to beat or win
over others. This should be transmuted into: strength
for the purpose of accomplishment. Learning to pace one's
self is also needful, as Pluto/Mars pushes to the point
of self depletion and the energies are exhausted in
useless efforts.

JUPITER - Unrealistic, unwarranted optimism is possible
through this aspect. The individuals so influenced must
learn that the negative experiences are as essential for
growth as those which are positive; unpleasant tasks or
problems must be met head-on. The tendency to vast accummulation of goods is present, as with the trines, but
in this case it is primarily for self. Pluto square Jupiter must learn to part with at least a few things here
and there or can find themselves swamped with bric-a-
brac. Unevolved Pluto/Jupiter is prone, too, to mistak-
ing his law for the law and should beware of making
harsh judgments on others. Generosity and fair judgments
are called for with this aspect.

SATURN - These people are called upon to work, and work
hard. Rarely if ever does the need diminish, since
these types are challenged to learn the lessons of precise, responsible effort put forth. Even in lavish
environments which ordinarily would not involve one in
the workaday world, the individual will still be found
carrying on with rather ordinary daily duties. The Pluto
square Saturn implies an evolutionary path which leads

toward the assumption of responsibility and the development of a patient, precise, dutiful life pattern.

URANUS - The Plutonian-Uranian has stepped aside from the mainstream in some aspect of his life and frequently "just to be different". These are people who are usually unfettered by heavy responsibilities and the issue is, can they learn to handle freedom? They are generally involved in off-beat activities, or participate in ordinary activities in an off-beat way, since they are anti-regulation and sometimes even anti-social. Extraordinarily impulsive, these people take giant leaps without looking and must learn to transmute this quest for new experience into Uranian inventiveness and originality which can be brought to the task at hand.

NEPTUNE - People represented by this aspect are supersensitive, "touchy", and capable of being first rate fanatics on some subject. They tend to bring too much life to their dreams instead of bringing the dreams to life, and could learn much about this subtle difference by observing the Pluto trine Neptune types. Much power and energy is directed, through Pluto, to the artistic and imaginative faculties; when channeled, this can be a tremendous force for betterment. These individuals, as with the trines, are capable of great sacrifice, but should watch for any tendency toward feelings of martyrdom or self pity. Idle fantasy must be transformed into artistry, self sensitivity into compassion.

****

78.

CONCLUSION

When we learn to express the finest qualities
of Pluto we will be able to annihilate negativities in our world, instead of each other...
to push past the barriers which divide people...
to use our Plutonian strength to dominate our
own weaknesses...to make universal goals such
as peace and brotherhood realities. FDR put it
well:

> "Our earth is but a small star
> in the great universe. Yet of
> it we can make, if we choose,
> a planet unvexed by war,
> untroubled by fear or hunger,
> undivided by senseless dis-
> tinctions of race, color or
> theory." *

This is the message of Pluto.

*Prayer, written for and read by Franklin
D. Roosevelt to the United Nations.

NOTES

Introduction
1. By Margaret Mead; Selected by Dee Danner Barwick. _Great Words of Our Time_, Hallmark Editions, n.d.

SECTION II

Pluto In Taurus
1. John Bartlett. _Familiar Quotations_, Boston-Toronto, 1955, p. 540
2. T. Harry Williams and The Editors of Life. _The Life History of the United States_, Vol. 5, New York, 1963, p. 13
3. _Ibid._, pp. 13-15
4. _Ibid._, pp. 15-17
5. _Ibid._, pp. 73-74
6. Editors of Time-Life Books. _This Fabulous Century_, 1870-1900, New York, 1971, p. 125
7. Ibid., p. 126
8. Williams, _op.cit._, p.9
9. Williams, _op.cit._, Chronology
10. Williams, _op.cit._, p. 86
11. Ernest R. May and the Editors of Life. _The Life History of the United States_, Vol. 9, New York, 1964, p. 7
12. Editors of Time-Life, TFC, 1870-1900, _op.cit._, p.186
13. Editors of Time-Life Books. _This Fabulous Century_, 1900-1910, New York, 1971, p.29
14. Editors of Time-Life, TFC, 1870-1900, _op.cit._, p.144
15. Editors of Time-Life, TFC, 1870-1900, _op.cit._, p.147
16. Editors of Time-Life, TFC, 1870-1900, _op.cit._, p.149
17. Bernard A. Weisberger and Editors of Life. _The Life History of the United States_, Vol. 7, New York, 1964, p.37
18. _Ibid._, pp. 32-37
19. _Ibid._, p. 156
20. Editors of Time-Life, TFC, 1870-1900, _op.cit._, pp. 44-45

Pluto In Gemini
1. Dee Danner Barwick. _op.cit._, p. 12
2. Bernard A. Weisberger and Editors of Life. _The Life History of the United States_, Vol. 8, New York, 1964, p. 131
3. _Ibid._, p. 131
4. _Ibid._, p. 132
5. _Ibid._, p. 127
6. _Ibid._, p. 133

7. Editors of Time-Life. TFC 1870-1900, op.cit.,p.112
8. Weisberger, op.cit., p.156
9. Weisberger, op.cit., p. 166
10. Editors of Time-Life. This Fabulous Century, 1900-1910, New York, 1971, pp. 228-231
11. Editors of Time-Life, TFC, 1870-1900, op.cit.,p.124
12. Compiled from: The Life History of the United States, Vols. 5 through 12, Chronology. The Evolution of the Machine by Ritchie Calder. Machines by Robert O'Brian and Editors of Life.
13. Ernest R. May and Editors of Life. The Life History of the United States, Vol. 9, New York, 1964, Cover
14. Editors of Time-Life, TFC, 1900-1910, op.cit.,p.29
15. Editors of Time-Life, TFC, 1900-1920, op.cit.,p.29
16. Ernest R. May, Vol. 9, op.cit., p.116
17. J.B. Priestley. The Edwardians, New York, 1970,passim
18. Weisberger, Vol. 8, op.cit., p. 60
19. Editors of Time-Life, TFC, 1870-1900, op.cit., p.78
20. Ernest R. May, Vol. 9, op.cit., Chronology
21. Weisberger, Vol. 7, op.cit., p.29
22. Editors of Time-Life, TFC, 1870-1900, op.cit., p.237
23. Editors of Time-Life, TFC, 1870-1900, op.cit., p.237
24. Weisberger, Vol. 7, op.cit., p. 45
25. Editors of Time-Life, TFC, 1870-1900, op.cit., p.142
26. Editors of Time-Life, TFC, 1870-1900, op.cit., p.142
27. Weisberger, Vol. 8, op.cit., p.156

Pluto In Cancer
1. Editors of American Heritage. The American Heritage History of the 1920's and 1930's, New York, 1970,p.25
2. Editors of Time-Life.This Fabulous Century, 1910-1920, New York, 1971, p.238
3. Ibid., p. 238
4. Ernest R. May and Editors of Life. The Life History of the United States, Vol. 10, New York, 1964, p.11
5. Editors of Time-Life, TFC, 1910-1920, op.cit.,p.230
6. Ernest R. May, Vol. 10, op.cit., p. 81
7. William E. Leuchtenburg and Editors of Life. The Life History of the United States, Vol. 11, New York, 1964, p. 29
8. Ibid., Chronology
9. Editors of Time-Life. This Fabulous Century, 1920-1930, New York, 1971, p. 23
10. Leuchtenburg, Vol. 11, op.cit., p. 8
11. Leuchtenburg, Vol. 11, op.cit., p. 78
12. Leuchtenburg, Vol. 11, op.cit., p. 9
13. Leuchtenburg, Vol, 11, op.cit., p. 10
14. Leuchtenburg, Vol. 11, op.cit., p. 11
15. Editors of Time-Life, TFC, 1920-1930, p. 123
16. Alan C. Collins. The Story of America, New York, 1953, p. 353

17. Editors of Time-Life, TFC, 1920-1930, <u>op</u>.<u>cit</u>.,p.175

Pluto In Leo
  1. John Bartlett. <u>Familiar</u> <u>Quotations</u>, Boston-Toronto,
     1955, p.919
  2. Editors of "Time" Magazine. <u>Time</u> <u>Capsule/1939</u>, New
     York, 1968, p. 77
  3. Editors of "Time" Magazine. <u>Time</u> <u>Capsule/1941</u>, New
     York, 1967, p. 160
  4. William E. Leuchtenburg and the Editors of Life.
     <u>The</u> <u>Life</u> <u>History</u> <u>of</u> <u>the</u> <u>United</u> <u>States</u>, Vol. 11,
     New York, 1964, p. 98
  5. Editors of "Time" Magazine. <u>Time</u> <u>Capsule/1940</u>,
     New York, 1968, p. 60
  6. Editors of"Time"Magazine, TC/1941, <u>op</u>.<u>cit</u>., pp.10-11
  7. Editors of Year. <u>Turbulent</u> <u>20th</u> <u>Century</u>, New York,
     1961, p.180
  8. Leuchtenburg, Vol. 11, <u>op</u>.<u>cit</u>., p. 153
  9. Dee Danner Barwick, <u>op</u>.<u>cit</u>., p. 22
 10. Editors of "Newsfront" Magazine. <u>1950-1960</u> –
     <u>Historic</u> <u>Decade</u>, New York, 1960, passim
 11. <u>Ibid</u>., p. 162
 12. Editors of Year. Historic Decade, <u>op</u>.<u>cit</u>., passim
     Luchtenburg, Vol. 11, <u>op</u>.<u>cit</u>., pp. 113-125
 13. Editors of Time-Life. <u>This</u> <u>Fabulous</u> <u>Century</u>,
     1940-1950, New York, 1971, p. 188

Pluto In Virgo
  1. Dee Danner Barwick, <u>op</u>.<u>cit</u>., p. 52
  2. Editors of Time-Life Books. <u>This</u> <u>Fabulous</u> <u>Century</u>,
     1960-1970, New York, 1972, p. 204
  3. <u>Ibid</u>., p. 206
  4. <u>Ibid</u>., p. 207
  5. <u>Ibid</u>., p. 205
  6. Editors of Year. <u>Historic</u> <u>Decade</u> <u>1950-1960</u>, New
     York, 1961, p. 168
  7. Ibid., Chronology
  8. <u>Ibid</u>., p. 168
  9. Editors of Time-Life Books, 1960-1970, <u>op</u>.<u>cit</u>.,p.266
 10. Barwick, <u>op</u>.<u>cit</u>., p. 26
 11. "Reaching Beyond The Rational" (article) <u>Time</u>
     <u>Magazine</u>, April 23, 1973, p. 86
 12. Editors of Time-Life Books, TFC, 1960-1970, <u>op</u>.<u>cit</u>.,
     p. 84
 13. Editors of Time-Life Books, TFC 1960-1970, <u>op</u>.<u>cit</u>.,
     p. 183
 14. L.H. Weston. <u>The</u> <u>Planet</u> <u>Vulcan</u>, <u>History</u>, <u>Nature</u>,
     <u>Tables</u>, Washington, D.C., n.d., p. 18

15. "The Wow Horse Races Into History" (article)
    Time Magazine, June 11, 1973, p. 85

Pluto In Libra
1. Barwick, op. cit., p.55

Pluto In Scorpio
1. Barwick, op.cit., p. 30

SECTION III

1. Barwick, op.cit., p. 24

BIBLIOGRAPHY

American Heritage Magazine Editors. The American Heritage History of the 1920's and 1930's, American Heritage Publishing Co., 1970, New York.

Associated Press, Editors of. The World in 1965, The World in 1966, The World in 1967, The World in 1968, The World in 1969, Pub. - Associated Press

Bacher, Elman. Studies in Astrology, Vol. II, Chapter VIII: "Pluto - Principle of Frozen Fire", The Rosicrucian Fellowship, Calif., 1962

Bartlett, John. Familiar Quotations, Little, Brown and Co. Boston-Toronto, 13th Ed. Rev., 1955

Barwick, Dee Danner. Great Words of Our Time, Hallmark Editions, n.d.

Benjamine, Elbert. The Influence of The Planet Pluto Including Ephemeris of Pluto, 1840 to 1960, Aries Press, Chicago, 1939

Brunhubner, Fritz. Pluto, American Federation of Astrologers, Washington, D.C., n.d.

Crummere, Maria Elise. The Age of Aquarius, Golden Press, New York, 1960

Davidson, Dr. W.M. "New Light On The Signs of the Zodiac", (mimeo), Reprint of 1950 Yearbook, American Federation of Astrologers, Washington, D.C.

DeVries, Leonard. Panorama 1840-1865, Houghton-Mifflin Co., Boston, 1969

Emerson, Edwin. History of the 19th Century - Year by Year, P.F. Collier & Son, New York, 1961

Gernsheim, Helmut and Alison. The Recording Eye, G.P. Putnam's Sons, New York, 1960

Grell, Paul R. Keywords, American Federation of Astrologers, Washington, D.C., 1970

Hutchins, Robert M., and Adler, Mortimer J. The Great Ideas Today - 1970, 1971, Encyclopaedia Britannica, Inc., Praeger Pub. Inc., New York-Washington, 1970, 1971

Jones, Marc Edmund. Astrology, How and Why It Works, David McKay, Philadelphia, 1945

Laver, James. Manners and Morals of The Age of Optimism, 1848-1914, Harper & Row, New York, 1966

Leuchtenburg, William E. and Editors of Life. The Life History of The United States, Vol.11 - New Deal and Global War, Vol. 12 - The Great Age of Change, Time, Inc., Book Div., New York, 1964

Luntz, Charles. Vocational Guidance by Astrology, Llewellyn Pub., St. Paul, 1962

May, Ernest R. and Editors of Life. The Life History of The United States, Vol. 9 - The Progressive Era, Vol. 10 - War, Boom and Bust, Time, Inc., Book Div., New York, 1964

Muir, Ada. "Pluto The Redeemer" (Booklet), Llewellyn Publications, St. Paul, 1967

Newsfront Editors. Year - 1971 Edition, New York, 1971

Priestley, J.B. The Edwardians, Harper & Row, New York, 1970

Pryor, Olive Adele. "Pluto - Cardinal or Angular" (article) AFA Bulletin, Vol. 33, #4, American Federation of Astrologers, Washington, May, 1971

Rosicrucian Fellowship, The. Ephemeris, 1961-1973, The Rosicrucian Fellowship, Calif.

Time-Life Books, Editors. This Fabulous Century - 1870-1900, 1900-1910, 1910-1920, 1920-1930, 1930-1940, 1940-1950, 1950-1960, 1960-1970, Time-Life Books, New York, 1971.

Time-Life Books, Editors. Time Capsule, 1939, 1940, 1942, 1943, 1945, Time-Life Books, New York, 1968

Time-Life Books, Editors. Time Capsule, 1968, Time-Life Books, New York, 1969

Time Magazine. Second Thoughts About Man IV, "Reaching Beyond The Rational" pp. 83-86 (article), Time Mag., April 23, 1973

Time Magazine. Cover Story, "The Wow Horse Races Into History" pp. 85-91, Time Mag., June 11, 1973

Weston, L.H. The Planet Vulcan, History, Nature, Tables, American Federation of Astrologers, Washington, D.C.,n.d.

Williams, T. Harry and Editors of Life. The Life History of the United States, Vol. 5 - The Union Sundered, Vol. 6 - The Union Restored, Vol. 7 - The Age of Steel and Steam, Vol. 8 - Reaching For Empire, Time, Inc., Book Div., New York, 1964

Year, Editors of. 1950-1960, Historic Decade, Year, New York, 1960

Year, Editors of. Turbulent 20th Century, Year, New York, 1961